STUDY GUIDE

to accompany

Jordan-Bychkov and Domosh's

THE HUMAN MOSAIC

Eighth Edition

by

Michael A. Kukral, Ph.D.
Ohio WesleyanUniversity

W.H. FREEMAN AND COMPANY
NEW YORK

Study Guide to accompany Jordan-Bychkov/Domosh's ***The Human Mosaic: A Thematic Introduction to Cultural Geography, Eighth Edition.***

ISBN: 0-7167-3802-3

Third printing, 1999

CONTENTS

Acknowledgments

About the Author

Preface

A Note to My Colleagues

To the Student: How To Use This Guide

Chapter 1: The Nature of Cultural Geography 1

Chapter 2: People on the Land 21

Chapter 3: The Agricultural World 39

Chapter 4: Political Patterns 57

Chapter 5: The Mosaic of Languages 75

Chapter 6: Religious Realms 93

Chapter 7: Folk Geography 111

Chapter 8: Popular Culture 125

Chapter 9: Ethnic Geography 141

Chapter 10: The City in Space and Time 159

Chapter 11: The Urban Mosaic 173

Chapter 12: Industrial Geography 189

ACKNOWLEDGEMENTS

Learning, teaching, experience and education have always been very important in my family. This brief work is dedicated to the great teachers in my life: my parents. To the memory of my late father, Clarence Ferdinand Kukral, an upholsterer in Cleveland for sixty-two years, and to my mother, Ada Mae Kukral, a wonderful artist from a farm in Bath Township, Ohio.

ABOUT THE AUTHOR

Dr. Michael A. Kukral is a professor of geography at Ohio Wesleyan University. He received his Ph.D. from the University of Kentucky and has been distinguished for teaching excellence at both Ohio University and the University of Kentucky. Dr. Kukral was a Fulbright Fellow in the Czech capital of Prague during the "Velvet Revolution" and is the author of Prague 1989: Theater of Revolution, A Study in Humanistic Political Geography. He has traveled extensively in continental Europe and has taught field study courses there, as well as in Wrightsville Beach, North Carolina. A native of the Cuyahoga Valley, Dr. Kukral is also co-owner of S & M Skates, Inc., and lives and writes in Athens, Ohio.

PREFACE

A Note to My Colleagues

There is no standard format for writing study guides. After perusing about a dozen guides and talking with many of my former students at the University of Kentucky, Ohio Wesleyan University, and Ohio University, I recognized two common problems that I seek to alleviate in this work.

First, many study guides replicate and summarize the textbook and are simply too lengthy (and thus an expensive addition to the textbook). Several students told me that when study guides are overly thorough, they find little reason to use or purchase the textbook. This can lead to serious problems in student comprehension of text and course material, especially given the highly visual nature of most texts in geography.

Secondly, some study guides designed for college introductory courses contain too many outside projects and assignments. The problem with this extraneous material is that it is rarely used by students. Most students in an introductory geography course are not geography majors (unfortunately) and already have enough trouble completing the assignments from their instructor. I have never met a student who has completed a study guide project on his or her own initiative.

My intention with this guide is to provide a simple and hopefully useful method for students to better understand the maps, concepts, and vocabulary of the textbook. Most students I questioned concluded that the most useful portions of any study guide were sections on key terms and the practice tests. I have expanded both of these sections from the length found in the average study guide. In addition, a new and innovative section on reading and interpreting selected maps from the textbook is included.

In summary, the value of any study guide is measured by its ability to help students better understand the textbook. Study guides cannot be written to cover other course material because non-text material is always included at the instructor's discretion. In the field of geography, as we all know, this material can be extremely diverse.

I personally welcome your comments, ideas, and especially criticisms of this study guide for the purpose of improving further editions. Please contact me directly by calling (740) 594-3670 or writing to: 27 Brown Avenue, Athens, Ohio 45701.

To the Student: How to Use this Guide

There is no standard method of studying that works for everyone, but there are many methods of studying that work for many people.
Some students study best either while listening to music, at the library, sitting outside, with a study partner, or in the privacy of their room. Despite the location you select, it is your study methods and techniques that will determine your ability to succeed in college. Using this Study Guide presents one such method of textbook comprehension and retention of text information.

Follow this plan of attack for using the Study Guide to Accompany the Human Mosaic:

1. **Peruse** the Human Mosaic before beginning your first reading assignment. This basically involves looking through the book, reading some brief sections that catch your attention, and becoming familiar with the organization of the book. You should note the sections of the book such as the Contents, Preface, Index, and most importantly, the Glossary. Be sure to use the glossary frequently

2. When you are given a reading assignment from the Human Mosaic, follow this logical pattern:

    a. Peruse the chapter to get an idea what it is about.

    b. Attentively read the chapter, perhaps in sections.

    c. Take notes, ask yourself questions, and perhaps highlight key statements while reading or re-reading the chapter. (Do not highlight everything that seems important or it all becomes ugly and meaningless!)

    d. Do NOT gloss over the maps, diagrams, and photographs. These are very important to geographers and they really summarize an enormous amount of information if you interpret them correctly, thoughtfully, and thoroughly.

    e. Now you may begin using the **Study Guide** to enhance your textbook readings. Complete all initial sections of the Study Guide chapter **before** taking the Practice Tests.

    f. Review the textbook chapter. Now you are ready to determine your progress by taking the Practice Tests. Do NOT look for the answers in the textbook while you are taking the tests. This will not help you (Its is like cheating at solitaire).

    g. Check your Practice Test answers with the answer key in the back of the Study Guide. Review your answer selections, especially the ones you missed, and learn from your mistakes. Rewrite your incorrect answers. Review the textbook chapter once more, rewrite your notes, get some sleep, and never miss class.

CHAPTER ONE

THE NATURE OF CULTURAL GEOGRAPHY

Extended Chapter Outline (including Key Terms)

I. Definition and Historic Development of **Geography**
   A. The growth and development of Geography
      1. Biographies: Humboldt and Ritter

II. What is **Cultural** Geography?
   A. learned collective behavior
   B. spatial variation and spatial pattern
   C. physical environments
   D. the example of wheat cultivation

III. Culture **Region**
   A. definition of Region and **Culture** Region
   B. Maps: the essential tool
   C. **Formal** Culture Regions
      1. core/periphery pattern
   D **Functional** Culture Regions
      1. nodes
   E. **Vernacular** Culture Regions
      1. a perceived region

IV. Cultural **Diffusion**
   A. definition: spatial spread of learned ideas
   B. independent invention
   C. expansion diffusion and relocation diffusion
      1. stimulus, hierarchical, and contagious diffusion
   D. time-distance decay
      1. absorbing and permeable barriers
   E. neighborhood effect
   F. biography: Tortsen Hagerstrand

V. Cultural **Ecology**
   A. definition: cause-and-effect interplay
   B. Environmental Determinism
   C. Possibilism
      1. cultural adaptation
      2. adaptive strategy

D. Environmental Perception
   1. the example of geomancy
   2. natural hazards and hazard zones
E. Humans as Modifiers of the Earth
   1. biography: George Perkins Marsh

VI. Cultural **Integration**
   A. definition: systemically and spatially intertwined
   B. cultural determinism
   C. Social Science
      1. scientific method
      2. economic determinism
      3. model building
   D. Humanistic Geography
      1. place and place meaning
      2. humanistic views and subjectivity
      3. biography: Anne Buttimer

VII. Cultural **Landscape**
   A. definition: humanized landscape
   B. Meitzen and diagnostic interpretation
   C. interpretations on landscape
   D. three principal aspects of cultural landscape
      1. settlement patterns
      2. land-division patterns
      3. architecture

VII. **Conclusion**
   A. example: the American log house
   B. suggested additional readings

LEARNING OBJECTIVES OF CHAPTER ONE

After reading this chapter <u>and</u> studying the maps and illustrations, you should be able to:

1. Briefly trace the historical development of Geography.

2. Define "geography" and how geographers approach the study of the Earth and its people.

3. Grasp the meaning of Cultural Geography and understand its basic topics, themes, and areas of study.

4. Describe, outline, compare, and give examples from the five major themes of cultural geography: culture region, cultural diffusion, cultural ecology, cultural integration, and cultural landscape.

5. Summarize the life and contributions of several important geographers, past and present.

6. Begin to use maps as essential tools of your education and understanding of the world.

SELECTED MAP READING AND INTERPRETATION

This section of the Study Guide is intended to heighten your map reading and interpretation skills. It will also help you apply the text readings to visual and spatial display of concepts, themes, and examples in cultural geography.

A world atlas will be very useful in completing this section of the Study Guide and will enhance your comprehension of the maps in the textbook. Ask your instructor to recommend an appropriate atlas to purchase (or visit the map collection at your library). A world atlas is essential for your personal reference library, not only during this course, but throughout your college career.

After reading the text and then studying the accompanying map and its captions, answer the following questions:

FIGURE 1.2 "Areas of wheat production in the world"

1. What is the specific single trait on which this map is based?

   ____

   ____

2. What is the scale of this map?

   a. local b. regional c. global d. continental

3. Are there any spatial patterns of wheat production evident at this scale? For example, is wheat cultivation stronger in the Tropics (between the Tropic of Cancer and the Tropic of Capricorn) or in the temperate latitudes (between the latitudes of 23 1/2 degrees and 66 1/2 degrees North or South)?

   ____

   ____

   ____

4. According to the map, where are the areas of significant wheat production in Africa, North America, and Russia?

   ____

   ____

   ____

FIGURE 1.4 "Two-trait formal culture regions of Europe"

1. What are the three major branches of Christianity portrayed on this map?

 ____________________, ____________________, ____________________

2. What are the three major language families on this map?

 ____________________, ____________________, ____________________

3. Explain what is shown and indicated by the yellow areas on this map:

 ____

 ____

 ____

4. State the two traits for the following countries:

 Spain: ____________________ and ____________________
 Poland: ____________________ and ____________________
 Norway: ____________________ and ____________________
 Austria: ____________________ and ____________________

5. What theme or themes in cultural geography does this map illustrate?

 ____

 ____

 ____

FIGURE 1.6 "Traditional rural formal cultural regions of North America"

1. Explain why southern Ohio and the coast of California are part of the same ("Midland") culture region:

 ____

 ____

 ____

2. What culture traits are found in the "Lower South?"

 ____

 ____

 ____

3. What are the two major "French trait" formal culture regions portrayed on the map?

 ____________________________ and ____________________________

FIGURE 1.8 "East vs. west in Germany"

1. Which two lines of cultural-historical demarcation roughly correspond to the former Iron Curtain?

 ____________________________ and ____________________________

2. Catholic majorities are located in Austria, ________________, and ________________.

3. What theme or themes in cultural geography does this map illustrate?

____

____

____

FIGURE 1.9 "Dixie: a vernacular region"

1. How does the vernacular region of "Dixie" correspond to the "Lower South" region in Figure 1.6?

____

____

____

2. Can you think of any cultural reasons why the use of the term "Dixie" is <u>not</u> as common in Virginia and Florida as in the "heart of Dixie"? Does this mean that Virginia and Florida share similar culture traits?

____

____

____

3. Explain how the information was collected to create this map:

____

____

____

4. Is this map a good example of core/periphery patterns or a better example of a region with nodes? Explain your reasoning.

____

____

____

FIGURE 1.11 "Early diffusion of the printing press in Europe"

1. Does the spatial pattern on this map more strongly suggest expansion or relocation diffusion (see Figure 1.10)? Explain the reasoning supporting your answer.

____

____

____

2. Is the idea of time-distance decay evident on this map? Why or why not?

____

____

____

3. Does this map illustrate the idea of neighborhood effect? Why or why not?

CREATE YOUR PERSONAL GLOSSARY OF KEY TERMS, PEOPLE, AND PLACES

In the space below, write a definition and provide an example of each key term that is sufficient for **your understanding**. It is an excellent study habit to organize your response in three parts:

1. a formal definition or identification from the textbook
2. a definition of the key term in **your own words**
3. an example to add greater understanding of the key term

KEY TERMS, PEOPLE, AND PLACES FROM CHAPTER ONE:

1. geography

____

____

____

2. spatial patterns

____

____

____

3. analytical geographic research

____

____

____

4. Alexander von Humboldt

____

____

____

5. Carl Ritter

____

____

____

6. physical geography

____

____

____

7. cultural geography

8. culture region

9. human traits

10. formal culture region

11. core/periphery pattern

12. functional culture region

13. cultural homogeneity

14. political borders

15. vernacular culture region

16. cultural diffusion

17. expansion diffusion

18. relocation diffusion

19. hierarchical diffusion

20. contagious diffusion

21. stimulus diffusion

22. time-distance decay

23. permeable barriers

24. neighborhood effect

25. cultural ecology

26. environmental determinism

27. possibilism in geography

28. cultural adaptation

29. environmental perception

30. geomancy

31. natural hazards

32. human-as-modifier theme

33. George Perkins Marsh

34. Judeo-Christian tradition

35. cultural integration

36. cultural determinism

37. economic determinism

38. model building

39. culture-specific model

40. humanistic geography

41. "place" and "space"

42. Anne Buttimer

43. post-modernism

____

____

____

44. cultural landscape

____

____

____

45. settlement forms

____

____

____

46. land-division patterns

____

____

____

47. traditional American log house

____

____

____

CHAPTER ONE REVIEW: Self-evaluation Tests

PART ONE: Multiple-choice

Circle the best answer for each question. When you are finished, read each question again with your selected answer. After you are satisfied with your practice test, use the Answer Key in the back of the Study Guide to check your responses.

1. According to Krishtalka and others, the foremost tool of geography is:
   a. observation
   b. a camera
   c. travel
   d. a map
   e. a compass

2. The word "geography" is a Greek word meaning literally:
   a. to study the earth
   b. to describe the earth
   c. to study various countries
   d. to write about the natural environment
   e. to analyze cultures of the world

3. During Europe's Dark Ages, the field of academic geography was taken over by the:
   a. Germans
   b. Chinese
   c. English
   d. Arabs
   e. French

4. In the 1700s, German scholars came to view geography as the study of:
   a. interrelated spatial patterns
   b. exploration and travel
   c. natural phenomena
   d. andscape interpretation
   e. ecosystems

5. The primary interest of Alexander von Humboldt was physical geography, while that of his associate Carl Ritter was:
   a. political geography
   b. economic geography
   c. cultural geography
   d. humanistic geography
   e. military geography

6. The authors of your textbook define culture as:
   a. spatial variation of humans
   b. inherited traits of humans
   c. traditional lifestyles
   d. learned human behavior
   e. human perception and actions

7. Cultural Geography is the study of the spatial functioning of society and:
   a. spatial variations among cultural groups
   b. the analysis of the physical environment
   c. human perception
   d. the built and natural landscape
   e. various behavioral traits of humans

8. A Culture Region is a geographical unit based in:
   a. vernacular characteristics
   b. traditional patterns
   c. human traits
   d. complex behavioral patterns
   e. physical properties

9. A French-language culture region is an example of a:
   a. functional culture region
   b. formal culture region
   c. vernacular culture region
   d. multiple-trait area
   e. none of the above

10. Formal regions in geography often display a core/periphery pattern.
   a. true
   b. false

11. Cultural homogeneity is the hallmark of a:
   a. functional culture region
   b. formal culture region
   c. vernacular culture region
   d. multiple-trait area
   e. none of the above

12. Nodes, or central points where activities are coordinated and directed, are a common characteristic of a:
   a. functional culture region
   b. formal culture region
   c. vernacular culture region
   d. multiple-trait area
   e. none of the above

13. Clearly defined borders are a common feature of a:
   a. functional culture region
   b. formal culture region
   c. vernacular culture region
   d. multiple-trait area
   e. none of the above

14. Which of the following is NOT an example of a functional culture region?
   a. coal fields of northern Appalachia
   b. newspaper circulation area
   c. Korean-language area
   d. Cleveland metropolitan area
   e. the state of Burma

15. A vernacular culture region is often considered to be a region:
   a. clearly demarcated on a map
   b. having strong functional and formal features
   c. perceived to exist by its inhabitants
   d. with traditional structural traits
   e. none of the above

16. The geographer best associated with humanistic geography is:
    a. Carl Ritter
    b. Hubert Wilhelm
    c. Alexander von Humboldt
    d. George Perkins Marsh
    e. Anne Buttimer

17. Torsten Hagerstrand is usually associated with his ideas concerning:
    a. vernacular landscape
    b. diffusion
    c. ethnic geography
    d. landscape interpretation
    e. mental maps and natural hazards

18. The wavelike spread of ideas primarily involves the concept of:
    a. stimulus diffusion
    b. hierarchical diffusion
    c. contagious diffusion
    d. absorbing barriers
    e. none of the above

19. The study of the interaction between culture and environment is termed:
    a. cultural ecology
    b. neighborhood effect
    c. environmental science
    d. humanistic geography
    e. biogeography

20. Environmental determinism was a common doctrine of belief among geographers during the:
    a. mid-nineteenth century
    b. 1950s and 1960s
    c. late eighteenth century
    d. early twentieth century
    e. 1970s

21. The belief that cultural heritage is at least as important as the physical environment in affecting human behavior is usually associated with:
    a. environmental determinism
    b. possibilism
    c. two-trait regions
    d. single-trait regions
    e. material landscape

22. Geomancy is:
    a. a traditional system of land-use planning
    b. geography based on mystical symbols
    c. a method of weather forecasting
    d. based primarily in Latin America
    e. none of the above

23. The theme of cultural integration, if improperly used, can lead the geographer to:
    a. social science
    b. pre-adaptive behavior
    c. cultural determinism
    d. geomancy
    e. possibilism

24. Model building, theory, and space, are terms and devices usually associated with the___________ approach in cultural geography.
    a. humanistic
    b. man as modifier of the earth
    c. cultural ecological
    d. landscape interpretation
    e. social science

25. Place, topophilia, sense of place, and subjectivity are terms and ideas usually associated with the _________________ approach in cultural geography.
    a. humanistic
    b. man as modifier of the earth
    c. cultural ecological
    d. landscape interpretation
    e. social science

26. Most geographical studies of the cultural landscape have focused on three principal aspects: land-division patterns, architecture, and ____________ .
    a. barn types
    b. settlement forms
    c. ethnicity
    d. food and drink
    e. spatial-temporal patterns of transportation

PART TWO: Short answer and fill in the blank (probable essay type questions)

27. The authors of the textbook are Jordan-Bychov and _______________ .

28. The practical aspects of geography first arose among the ancient Greeks, Romans, _________________ and ___________________.

29. Whereas Alexander von Humboldt is usually associated with ________________ geography, Carl Ritter was often considered a father of modern ____________________ geography.

30. Cultural geography is the study of:

____

____

____.

31. The five geographical themes in cultural geography are:

a. ________________
b. ________________
c. ________________
d. ________________
e. ________________

32. A culture region is a

____

____

____.

33. Explain the difference between formal culture regions and functional culture regions:

____

____

____

____

34. Give an example of a core/periphery pattern:

____

____

____

35. Based on the maps in this chapter, in which culture region of North America do you live?

____

36. Explain why the German-language region is different from the country of Germany in the perspective of cultural geography:

____

____

____

37. How are vernacular culture regions different from formal and functional regions?

____

____

____

38. Explain the difference between hierarchical diffusion and stimulus diffusion:

39. What is meant by time-distance decay?

40. Explain the difference between environmental determinism, cultural ecology, and possibilism:

41. What is humanistic geography?

42. Examples of natural hazards are:

43. What role does 'perception' play in studies of cultural geography?

44. List some of the contributions of George Perkins Marsh to the field of geography:

45. Describe the theme of cultural integration:

46. Explain the difference between the geographic concepts of 'space' and 'place':

47. In your own words, give an example of 'placelessness':

48. What are culture-specific models?

49. List some actual examples of the cultural landscape:

CHAPTER TWO

PEOPLE ON THE LAND

Extended Chapter Outline (including Key Terms)

I. Demographic Regions
- A. Population Distribution
    1. population density
    2. ten most populous countries
    3. thickly, moderately, and thinly settled areas
    4. absolute and physiological density
- B. Patterns of Natality
    1. birthrates
    2. fertility rates
- C. The Geography of Mortality
    1. death rates
- D. The Population Explosion
    1. crucial elements
    2. geometric progression
    3. Malthus
    4. demographic stabilization
- E. Demographic Transformation
- F. Age Distributions
    1. population pyramid
- G. Geography of Gender
    1. sex ratio
    2. gendered spaces
- H. Standard of Living

II. Diffusion in Population Geography
- A. Migration
    1. voluntary and forced migration
    2. push-and-pull factors
- B. Diffusion of Fertility Control
    1. cultural influences
- C. Disease Diffusion
    1. example of AIDS

III. Population Ecology
- A. Environmental Influence
    1. climatic factors
    2. coastal locations

B. Environmental Perception and Population Distribution
    1. interregional migration
C. Population Density and Environmental Alteration
    1. changes in vegetation

IV. Cultural Integration and Population Patterns
    A. Cultural Factors
        1. food, crops, attitudes
        2. influence of gender
        3. personal space
    B. Political Factors
        1. government actions
    C. Economic Factors

V. The Settlement Landscape
    A. Farm Villages
        1. clustered settlements
            a. village pattern types
        2. environmental determinants
    B. Isolated Farmsteads
        1. global locations
    C. Semiclustered Rural Settlement
        1. village pattern types
    D. Reading the Cultural Landscape
        1. example of Spanish and Mayan influence

VI. Conclusion

LEARNING OBJECTIVES OF CHAPTER TWO

After reading this chapter <u>and</u> studying the maps and illustrations, you should be able to:

1. Understand the terms and concepts used by geographers to study the human population.

2. Give examples of the various factors influencing human migration and population diffusion.

3. Distinguish and discuss the differences among population distribution, population density, and population composition.

4. Show the relationship between population and the natural environment, including factors of human perception.

5. Understand the role of political, economic, and especially cultural phenomena in influencing the human population.

6. Distinguish various settlement patterns and traditional village types, such as the 'street village.'

7. Continue to improve your ability to read and interpret the cultural landscape.

SELECTED MAP READING AND INTERPRETATION

This section of the Study Guide is intended to heighten your map reading and interpretation skills. It will also help you apply the text readings to visual and spatial display of concepts, themes, and examples in cultural geography.

A world atlas will be very useful in completing this section of the Study Guide and will enhance your comprehension of the maps in the textbook. Ask your instructor to recommend an appropriate atlas to purchase (or visit the map collection at your library). A world atlas is essential for your personal reference library, not only during this course, but throughout your college career.

After reading the text and then studying the accompanying map and its captions, answer the following questions:

FIGURE 2.1 "Population density of the world"

1. It is well-known that China has the largest population in the world. The pattern of population distribution in China, however, shows a clear East-West division. Can you think of any environmental or economic factors influencing this pattern in China? Secondly, do you see a similar distribution pattern in the United States, Brazil, and Australia? Provide a few statements to support your answer.

   ____
   ____
   ____
   ____

2. Is it reasonable to state that population density is usually highest in coastal regions? Select a few specific countries or regions to support your answer.

   ____
   ____
   ____

3. What is the pattern of population density in the tropics? Use the continents of Africa, Asia, and South America in your answer.

   ____
   ____
   ____

4. Can you think of any reasons for the band of high population density running across the interior of Central Europe?

   ____
   ____
   ____

FIGURE 2.3 "The annual birthrate per thousand inhabitants"

1. What regions (East, West, North, Equatorial, Southern) of Africa show the highest birthrates?

2. Is there a clear North-South pattern present at the global scale?

3. What are the possible cultural influences affecting the birthrates in Australia and New Zealand?

FIGURE 2.4 "The total fertility rate (TFR) in the world"

1. Compare this map to Figure 2.3. What regions of the world have both high birthrates and a high TFR?

2. Can you think of any reasons why some Latin American countries, such as Honduras, have higher TFR than others?

3. Compare the data on this map with the data shown in Figure 2.1. Does this comparison illustrate that the population of China and India must be declining? Why or why not?

FIGURE 2.10 "Geography of contraception in the modern world"

1. Where do the highest percentages occur? Are these rich or poor areas?

2. The white areas on this map are an indication of what?

3. Compare this map with Figure 2.4. What can you interpret from this comparison?

____

____

____

<u>FIGURE 2.25 "Protein malnutrition, vegetarianism, and rice consumption in India"</u>

1. Do these maps show a relationship between vegetarianism and malnutrition in India? Why or why not?

____

____

2. Is there an evident relationship between predominantly rice consuming areas and malnutrition?

____

____

3. Can you think of any cultural, economic, or environmental reasons for the patterns on these maps?

____

____

____

CREATE YOUR PERSONAL GLOSSARY OF KEY TERMS, PEOPLE, AND PLACES

In the space below, write a definition and provide an example of each key term that is sufficient for **your understanding**. It is an excellent study habit to organize your response in three parts:

1. a formal definition or identification from the textbook
2. a definition of the key term in **your own words**
3. an example to add greater understanding of the key term

KEY TERMS, PEOPLE, AND PLACES FROM CHAPTER TWO:

1. demography

____
____
____

2. population density

____
____
____

3. demographic regions

____
____
____

4. population distribution

____
____
____

5. thickly, moderately, and thinly settled areas

____
____
____

6. physiological density

____
____
____

7. absolute density

8. natality

9. birthrates

10. fertility rate

11. mortality

12. death rates

13. population explosion

14. Thomas Malthus

15. demographic stabilization

16. demographic transformation

17. post-industrial period

18. population pyramid

19. sex ratio

20. gendered spaces

21. migration

22. voluntary and forced migration

23. push-and-pull factors

24. diffusion of fertility control

25. disease diffusion

26. AIDS, HIV-1, HIV-2

27. population ecology

28. preadaptation

29. sustainability

30. European coal fields

31. interregional migration

32. kitchen gardens

33. matrilineal societies

34. personal space

35. primogeniture

36. settlement landscape

37. clustered settlements

38. farm villages

39. street village

40. green village

41. strong-point and wet-point settlements

42. isolated farmsteads

43. semiclustered rural settlement

44. hamlet

45. loose irregular and row villages

46. Spanish-influenced architecture

47. cenote

48. Malthusian theory

49. population geography

CHAPTER TWO REVIEW: Self-evaluation Tests

PART ONE: Multiple-choice

Circle the best answer for each question. When you are finished, read each question again with your selected answer. After you are satisfied with your practice test, use the Answer Key in the back of the Study Guide to check your responses.

1. The current population of the Earth is about:
   a. 2.9 billion
   b. 9.9 billion
   c. 4.5 billion
   d. 12 billion
   e. 5.7 billion

2. Population geographers study the spatial and ___________ aspects of demography.
   a. physical
   b. physiological
   c. ecological
   d. cultural
   e. distributive

3. The five most populous countries are Brazil, United States, China, India, and:
   a. Russia
   b. Indonesia
   c. Nigeria
   d. Germany
   e. Pakistan

4. Less than five percent of the world's population lives in the United States.
   a. true
   b. false

5. 'Moderately settled areas' have about _________ to _________ persons per square mile.
   a. 500-700
   b. 250-500
   c. 60-250
   d. 2-60
   e. none of the above

6. The density beyond which people cease to be nutritionally self-sufficient is called:
   a. anatomical
   b. physiological
   c. nutritional
   d. biological
   e. caloric

7. Birth and death rates are measured in number per:
   a. hundred
   b. thousand
   c. hundred-thousand
   d. million
   e. square mile

8. The three main population clusters of the world are eastern Asia, the Indian subcontinent, and:
   a. Europe
   b. China
   c. eastern USA
   d. West Africa
   e. East Africa

9. At present rates, how many years will it take for the world population to double?
   a. 15
   b. 20
   c. 30
   d. 40
   e. 50

10. The American tropics, North Africa, the Middle East, and Central Asia all have __________ death rates.
    a. very high
    b. high
    c. low
    d. average
    e. very low

11. Rapid population growth among humans began around the year:
    a. 1450
    b. 1930
    c. 1890
    d. 1820
    e. 1700

12. Malthus stated that famine and __________ would be constant checks of population growth.
    a. natural disasters
    b. poverty
    c. disease
    d. war
    e. accidents

13. In preindustrial societies, birth and death rates are both normally:
   a. high
   b. low
   c. moderate
   d. very low
   e. none of the above

14. Recently settled areas typically have more males than females.
   a. true
   b. false

15. Living in a farm village places the family next to their land.
   a. true
   b. false

16. The most common type of migration is:
   a. forced
   b. push-and-pull
   c. international
   d. war refugee
   e. voluntary

17. The largest migration in history took place in the _________ century when more than 50 million people left Europe.
   a. twentieth
   b. nineteenth
   c. eighteenth
   d. seventeenth
   e. sixteenth

18. In Africa, AIDS is a disease primarily affecting the ____________ community.
   a. elderly
   b. upper economic
   c. heterosexual
   d. homosexual
   e. Islamic

19. Areas of extremely high heat or cold are considered to be _______________ climates from the perspective of humans.
   a. 'golden mean'
   b. wonderful
   c. defective
   d. marginal
   e. uninhabitable

20. Americans account for about ______ percent of the world's resources consumed each year.
   a. 5
   b. 10
   c. 15
   d. 20
   e. 40

21. The inheritance of all land by the firstborn son is termed:
   a. primogeniture
   b. Napoleonic Law
   c. Roman Law
   d. patrilineal
   e. matrilineal

22. Farm villages are common throughout much of the rural American Midwest.
   a. true
   b. false

23. The 'street village' is particularly common to:
   a. China
   b. Russia
   c. Latin America
   d. East Africa
   e. Southeast Asia

24. Before the Spanish conquest, most Mayans lived in ____________ villages.
   a. highland
   b. coastal
   c. isolated
   d. wet-point
   e. common

25. A 'cenote' is a:
   a. village public hall
   b. public cooking area
   c. skate wheel
   d. sinkhole
   e. village elder

PART TWO: Short answer (probable essay type questions)

26. What do population geographers study <u>and</u> what topics and processes are included in these studies?

____

____

____

27. Explain the difference between population distribution and population density:

28. Define the terms physiological density and absolute density:

29. How is the fertility rate measured and what can this statistic indicate in regard to population patterns?

30. What is meant by 'population explosion' and how and why did this event come about?

31. What are the contributions of Thomas Malthus to demographic studies? Are his ideas valid today?

32. Briefly define and categorize the demographic transformation:

33. What is a population pyramid and what are the important contributions they provide to demographers?

34. Describe several important push-and-pull factors of migration:

35. Provide several examples of forced migration:

36. Briefly describe the diffusion theory of AIDS:

37. What is population ecology?

38. Give some examples of environmental influence on population:

39. How does human perception play a role in distribution and settlement patterns?

40. Label and identify the various types of farm villages. How do isolated farmsteads differ?

CHAPTER THREE

THE AGRICULTURAL WORLD

Extended Chapter Outline (including Key Terms)

I. Agricultural Regions
   A. Shifting Cultivation
   B. Paddy Rice Farming
   C. Peasant Grain, Root, and Livestock Farming
   D. Mediterranean Agriculture
   E. Nomadic Herding
   F. Plantation Agriculture
   G. Market Gardening
   H. Commercial Livestock Fattening
   I. Commercial Grain Farming
   J. Commercial Dairying
   K. Livestock Ranching
   L. Nonagricultural Areas

II. Agricultural Diffusion
   A. Origin and Diffusion of Plant Domestication
   B. Origin and Diffusion of Animal Domestication
   C. Modern Innovations in Agriculture

III. Agricultural Ecology
   A. Cultural Adaptation
   B. Agriculturalists as Modifiers of the Environment
   C. Environmental Perception by Agriculturalists

IV. Cultural Integration in Agriculture
   A. Dietary Preferences
   B. Intensity of Land Use
   C. Amish Agriculture

V. Agricultural Landscapes
   A. Survey, Cadastral, and Field Patterns
   B. Fencing
   C. Hedging

VI. Conclusion and Suggested Readings

LEARNING OBJECTIVES OF CHAPTER THREE

After reading this chapter and studying the maps and illustrations, you should be able to:

1. Describe various forms of agriculture such as subsistence agriculture and commercial livestock fattening.

2. Trace the origin and places of plant and animal domestication.

3. Discuss the contributions of several important scholars and geographers, most notably Sauer, Meitzen, and von Thunen.

4. Explain and understand the diffusion of agriculture from one region or continent to another.

5. Describe the influence of the environment upon agricultural practices and how humans modify the earth to suit their agricultural needs and methods.

6. Understand and explain the role of human and environmental perception upon land use and agricultural production.

7. Discuss the effects of culture on dietary preferences.

8. Provide examples from around the world of various agricultural lifestyles and practices, including those of the Dasanetch.

9. Understand the various techniques of land survey, land division, fencing, and hedging that shapes the cultural landscape.

10. Continue to sharpen your skills in map reading and interpretation.

SELECTED MAP READING AND INTERPRETATION

This section of the study guide is intended to heighten your map reading and interpretation skills. It will also help you apply the text readings to visual and spatial display of concepts, themes, and examples in cultural geography.

A world Atlas will be very useful in completing this section of the study guide and will enhance your comprehension of the maps in the textbook. Ask your instructor to recommend an appropriate atlas to purchase (or visit the map collection at your library). A world atlas is essential for your personal reference library, not only during this course, but throughout your college career.

After reading the text and then studying the accompanying map and its captions, answer the following questions:

FIGURE 3.1 "Agricultural regions of the world"

1. What is the agricultural region stretching from France to Siberia and in what other areas of the world is it found?

____

____

____

2. Identify several regions and countries that are dominated by Livestock fattening:

____

____

____

3. In what regions of the world is Plantation agriculture practiced? Can you think of any reason why?

____

____

____

4. What agricultural regions are shared by Peru, Burkina Faso, Iraq, and North Korea?

____

____

FIGURE 3.12 "The origin and diffusion of agriculture"

1. What are the reasons why this map should be regarded as theoretical or speculative?

____

____

____

2. What is the historic name of the region acknowledged as the hearth area of the earliest plant and animal domestication? What are the present day countries of this Middle Eastern region?

3. Where is the region of domestication of vegetatively reproducing plants in the Americas? What vegetables were domesticated here and what was their possible path of diffusion?

FIGURE 3.14 "The diffusion of hybrid maize in the United States"

1. Based on the readings, can you explain the barriers this innovation encountered in New England, the South, and the Great Plains?

FIGURE 3.20 "Chemical dependence in agriculture, as exemplified by chemical fertilizers"

1. Which global regions and major countries are indicated as having 'extreme dependence'?

2. Can you think of any reasons why traditionally poorer countries of the world, such as Egypt, Suriname, and China are shown as extreme or heavy users?

3. Can you find and explain any regional patterns or relationships between this map and Figure 3.1?

FIGURE 3.22 "Ideal and actual distribution of types of agriculture in Uruguay"

1. In what ways does the spatial pattern of Uruguayan agriculture conform to von Thunen's model?

2. How is the pattern of agriculture different from von Thunen's model and what might cause the anomalies?

____

____

____

CREATE YOUR PERSONAL GLOSSARY OF KEY TERMS, PEOPLE, AND PLACES

In the space below, write a definition and provide an example of each key term that is sufficient for **your understanding**. It is an excellent study habit to organize your response in three parts:

1. a formal definition or identification from the textbook
2. a definition of the key term in **your own words**
3. an example to add greater understanding of the key term

KEY TERMS, PEOPLE, AND PLACES FROM CHAPTER THREE:

1. agriculture

____

____

____

2. agricultural regions

____

____

____

3. shifting cultivation

____

____

____

4. slash-and-burn agriculture

____

____

____

5. intertillage

____

____

____

6. subsistence agriculture

____

____

____

7. paddy rice farming

8. double-cropping

9. green revolution

10. Mediterranean agriculture

11. nomadic herding

12. plantation agriculture

13. luxury crops

14. neo-plantation

15. market gardening

16. Corn Belt

17. feedlots and commercial livestock fattening

18. commercial grain farming

19. agribusiness

20. commercial dairying and livestock ranching

21. hunting and gathering groups

22. 'buffalo commons'

23. environmental perception by agriculturalists

24. dietary preferences

25. agricultural productivity per capita

26. Johann Heinrich von Thunen

27. Isolated State model

28. concentric zone model

29. Dasanetch and also Tatar agriculture

30. August Meitzen

31. cadastral pattern

32. survey patterns

33. fragmented landholding

34. American rectangular survey system

35. metes and bounds surveying

36. 'sloughs'

37. open-field areas

38. hedgerow country

39. domesticated plant

40. Carl Sauer

41. Fertile Crescent

42. Mesoamerican crop complex

43. domesticated animal

44. 'gene banks'

45. agricultural adaptive strategies

46. cultural adaption (agriculturalist)

47. modification of the environment by agriculturalists

48. desertification

49. The Dust Bowl

50. cultural diffusion of the potato in Germany

51. cattle among the Dasanetch and cultural integration

52. threshing

____

____

____

53. stone or rock fence

____

____

____

CHAPTER THREE REVIEW: Self-evaluation Tests

PART ONE: Multiple-choice

Circle the best answer for each question. When you are finished, read each question again with your selected answer. After you are satisfied with your practice test, use the Answer Key in the back of the Study guide to check your responses.

1. About _____ percent of the world's working population is employed in agriculture.
   a. 2
   b. 5
   c. 15
   d. 30
   e. 45

2. In the United States the number of people involved in farming is less than _____ percent.
   a. 2
   b. 5
   c. 10
   d. 12
   e. 15

3. The agricultural practice of shifting cultivation is essentially:
   a. 'slash-and-burn' agriculture
   b. land-rotation system
   c. found primarily in Russia and other former communist states
   d. subsistence agriculture
   e. none of the above

4. The hallmark of paddy rice farming is:
   a. Irish stone fences
   b. environmental deterioration
   c. high protein productivity
   d. terraced paddy fields
   e. the water buffalo

5. Planting and harvesting the same land two or three times a year is known as:
   a. intertillage
   b. crop-rotation
   c. double-cropping
   d. biannual agriculture
   e. none of the above

6. The "green revolution" involved the introduction of:
   a. sugar cane
   b. chemical fertilizers
   c. organic methods
   d. traditional practices of sustainability
   e. gene banks

7. Llamas and alpacas serve as traditional beasts of burden in:
   a. South America
   b. Mexico
   c. Middle East and North Africa
   d. Southeast Asia
   e. Southeast Ohio

8. Traditional Mediterranean agriculture is based in part on the cultivation of primarily barley and:
   a. oats
   b. millet
   c. rye
   d. wheat
   e. rice

9. Nomadic herding is normally practiced in what type of environment?
   a. fertile valley floor
   b. tropical rain forest
   c. mountains or desert
   d. coastal plains
   e. temperate forests

10. Plantation agriculture usually maximizes the production of ____________ for Europeans and Americans.
   a. coffee and tea
   b. grains
   c. cattle and other livestock
   d. luxury crops
   e. fruits and vegetables

11. Crops which tend to grow best in tropical highland regions are:
    a. coffee and tea
    b. grains
    c. luxury crops
    d. fruits and vegetables
    e. none of the above

12. Market gardening is also referred to as:
    a. neo-plantation
    b. greenhouse farming
    c. truck farming
    d. market plantation
    e. organic farming

13. One of the most highly developed areas of commercial livestock fattening is:
    a. Alpine Europe
    b. Mediterranean region
    c. Northeastern United States
    d. Corn Belt
    e. Mesoamerica

14. Fattening livestock, especially cattle, is an efficient method of protein production.
    a. true
    b. false

15. The major wheat producing countries (producing 35 percent of the world's wheat) are the United States, Russia, Canada, Argentina, Kazakhstan and:
    a. Germany
    b. Australia
    c. India
    d. Italy
    e. Ukraine

16. Dairy farms located near large urban areas usually produce:
    a. butter
    b. cheese
    c. yogurt
    d. processed milk
    e. none of the above

17. Australia, New Zealand, South Africa, and Argentina produce 70 percent of the world's export:
    a. beef
    b. wool
    c. coffee and tea
    d. pork
    e. fruit crops

18. Geographer Carl Sauer believed that plant domestication developed as a response to hunger.
    a. true
    b. false

19. Perhaps the oldest primary region of agriculture is:
    a. Southeast Asia
    b. northwestern India
    c. the Fertile Crescent
    d. Mesoamerica
    e. northeastern China

20. The diffusion of many plant crops from the Western Hemisphere was accomplished by colonial powers, such as the Dutch and Portuguese.
    a. true
    b. false

21. What was developed to preserve what remains of domesticated plant variety?
    a. hybrid seeds
    b. the "green revolution"
    c. gene banks
    d. agriculture exclusion zones
    e. wild flower and plant nature preserves

22. A possible result of overgrazing of grasslands is:
    a. deforestation
    b. soil saturation
    c. desertification
    d. increased protein production
    e. none of the above

23. The Sahel is a critical region that is located:
    a. in southern India
    b. in the Middle East
    c. just south of the Sahara Desert
    d. in southern Africa
    e. in the interior of Australia

24. Above all, most farmers rely upon:
   a. climatic stability
   b. soil stability
   c. access to water
   d. temperature variation
   e. low labor costs

25. In von Thunen's isolated state model, the intensity of cultivation for any given crop increases with increasing distance from the market.
   a. true
   b. false

PART TWO: Short answer (probable essay type questions)

26. Briefly summarize the modified version of von Thunen's isolated state model:

27. State the principle contributions of the geographer August Meitzen:

28. Explain the difference between the American rectangular survey system and the "metes and bounds" survey system:

29. What are the reasons for establishing "long-lot" farms <u>and</u> where are examples of these farms found?

30. Briefly describe the example of the cultural diffusion of the potato in Germany.

31. What are the important themes are Carl Sauer's research, especially those concerning the domestication of plants and animals?

32. Explain and describe the highly distinctive type of subsistence agriculture called paddy rice farming.

33. State some of the problems as well as benefits associated with the "green revolution."

34. Briefly describe the problems of commercial livestock fattening in regards to nutritional efficiency.

35. What is the difference between a "suitcase farm" and agribusiness?

36. What are the major world regions of commercial dairying?

37. Describe the conditions and setting in which agriculture first arose.

38. What are some of the major crops domesticated by the native peoples of the Americas?

39. Briefly discuss the theme of cultural adaptation in agricultural geography.

CHAPTER FOUR

POLITICAL PATTERNS

Extended Chapter Outline (including Key Terms)

I. Political Culture Regions
   A. Independent Countries
      1. territoriality
      2. boundaries
      3. centrifugal and centripetal forces
   B. Supranational Political Bodies
   C. Electoral Geography

II. Political Diffusion
   A. Country Building as Diffusion
   B. Diffusion of Insurgencies and Innovations

III. Political Ecology
   A. Folk Fortresses
   B. The Heartland Theory
      1. Halford Mackinder
   C. Warfare and Environmental Destruction

IV. Politico-Cultural Integration
   A. The Nation-State
   B. The Multinational Country
   C. Ethnic Separatism
   D. The Cleavage Model
   E. Political Imprint on Economic Geography

V. Political Landscapes
   A. Imprint on the Legal Code
      1. border landscapes
   B. Physical Properties of Boundaries
   C. The Impress of Central Authority
   D. National Iconography on the Landscape

VI. Conclusion

LEARNING OBJECTIVES OF CHAPTER FOUR

After reading this chapter <u>and</u> studying the maps and illustrations, you should be able to:

1. Describe various types of states, countries, and other political bodies.

2. Understand the themes and concepts associated with country building.

3. Explain the political ecology of folk fortresses, warfare and environmental destruction, and Mackinder's heartland theory.

4. Explain the relationship of ethnicity and territory to the nation-state and the multinational state.

5. Describe the effects of political decision making on the cultural landscape.

6. Provide examples of the basic elements of the political landscape such as the boundary.

7. Better understand many of the world's current political conflicts and wars based on your knowledge of political and cultural geography.

SELECTED MAP READING AND INTERPRETATION

This section of the Study Guide is intended to heighten your map reading and interpretation skills. It will also help you apply the text readings to visual and spatial display of concepts, themes, and examples in cultural geography.

A world atlas will be very useful in completing this section of the Study Guide and will enhance your comprehension of the maps in the textbook. Ask your instructor to recommend an appropriate atlas to purchase (or visit the map collection at your library). A world atlas is essential for your personal reference library, not only during this course, but throughout your college career.

After reading the text and then studying the accompanying map and its captions, answer the following questions:

FIGURE 4.1 "The independent countries of the world"

1. The United States basically shares a border with only two countries. How many countries now border Russia and what are they? (don't forget Norway!)

    ____

    ____

    ____

2. What are the newly formed independent countries of Central Asia and the Caucasus region that were formerly part of the Soviet Union?

    ____

    ____

    ____

3. What independent countries were recently derived from the break-up of Yugoslavia, Ethiopia, and Czechoslovakia?

    ____

    ____

    ____

4. Locate these independent countries on the world map:
    Bhutan, Sierra Leone, Oman, French Guiana, Slovenia, Lithuania, Belize, Papua New Guinea, Swaziland, Haiti, and Vanuatu.

    ____

    ____

    ____

    ____

FIGURE 4.3 "Two independent countries, A and B"

1. These are real countries. Using an atlas, identify them.

2. The conditions shown and described here are for 1991. What was the territorial outcome of this dispute?

3. What and where are the actual religions, languages, and ethnicity of the various territories shown on this map?

FIGURE 4.5 "Some supranational political organizations in the Eastern Hemisphere"

1. Can you identify the reasons why certain European countries, such as Norway, Switzerland, Poland, and Austria, are not part of the European Community?

2. If British colonialism has basically ended, why are certain countries such as Zimbabwe and India part of the British Commonwealth?

3. What territories of the former Soviet Union are not part of the Commonwealth of Independent States? Why not?

FIGURE 4.10 "Russia developed from a core area"

1. Can you think of reasons why expansion to the east was greater than to the west? What were the barriers?

2. What environmental goals might have motivated Russian expansion?

3. What were the causes behind the recent contraction of the country? Do you foresee any further contraction?

FIGURE 4.11 "Diffusion of the Palestinian Arab uprising, or intifada, in the Israeli-ruled West Bank"

1. What specific types of diffusion appear to be involved?

2. What barriers kept the intifada from diffusing into the eastern region of the West Bank?

3. What is political and territorial history of Israel's illegal occupation of the West Bank? What is the position of the United Nations on this occupation?

4. In addition to the West Bank, what other Arab lands (including part of Lebanon) are ruled and occupied by Israel?

FIGURE 4.18 "Nation-states, multinational countries, and other types"

1. Identify ten nation-states from this map:

2. Why are the countries of Canada and Spain classified as old multinational states?

____

____

____

3. What is your opinion on the United States as a country evolving toward nation-statehood?

____

____

____

4. This classification is arbitrary and debatable. How would you change it, and why?

____

____

____

<u>FIGURE 4.26 "International political boundaries can influence transportation networks"</u>

1. Can you think of historical reasons from the political geography of this area for the pattern of motorable roads?

____

____

____

2. Do you think a similar pattern of motorable roads exist between Austria and Germany? What future events may cause the patterns on this map to change?

____

____

____

CREATE YOUR PERSONAL GLOSSARY OF KEY TERMS, PEOPLE, AND PLACES

In the space below, write a definition and provide an example of each key term that is sufficient for **your understanding**. It is an excellent study habit to organize your response in three parts:

1. a formal definition or identification from the textbook
2. a definition of the key term in **your own words**
3. an example to add greater understanding of the key term

KEY TERMS, PEOPLE, AND PLACES FROM CHAPTER FOUR:

1. political geography

2. independent countries

3. political culture region

4. territoriality

5. nationalism

6. national territory

7. enclave and exclave

8. "shoestring" countries

9. boundaries

10. marchlands

11. buffer state

12. satellite state

13. natural boundaries

14. ethnographic boundaries

15. geometric boundaries

16. unitary and federal spatial organization

17. centripetal forces

18. centrifugal forces

19. *raison d'etre*

20. supranational organizations

21. electoral geography

22. cleavages

23. gerrymandering

24. political diffusion

25. secession movements

26. diffusion of insurgencies and innovations

27. political ecology

28. folk fortress

29. manifest destiny

30. heartland theory

31. Halford Mackinder

32. rimland

33. politico-cultural integration

34. nation-state

35. multinational country

36. ethnic separatism

37. cleavage model

38. Kurdistan

39. political imprint

40. political landscape

41. imprint of the legal code

42. border landscapes

43. impress of central authority

44. "military landscapes"

45. national iconography

46. territorial imperative

47. The "West Bank"

48. Great Wall of China

49. Mount Rushmore

CHAPTER FOUR REVIEW: Self-evaluation Tests

PART ONE: Multiple-choice

Circle the best answer for each question. When you are finished, read each question again with your selected answer. After you are satisfied with your practice test, use the Answer Key in the back of the Study Guide to check your responses.

1. The Earth's surface is divided into approximately _____ independent countries.
   a. 230
   b. 320
   c. 280
   d. 210
   e. 190

2. Europe and Africa each have about the same number of independent countries.
   a. true
   b. false

3. Theoretically, the most desirable shape for a country is:
   a. elongated
   b. square
   c. hexagonal
   d. triangular
   e. none of the above

4. These areas are pieces of national territory separated from the main body of a country by the territory of another.
   a. peninsulas
   b. exclaves
   c. enclaves
   d. colonies
   e. protectorates

5. Until quite recently, many boundaries were not clear or sharp, but undefined, somewhat fuzzy zones called:
   a. buffer states
   b. frontiers
   c. hinterlands
   d. marchlands
   e. international zones

6. Mongolia and Nepal serve as good examples of:
   a. "shoestring" countries
   b. buffer states
   c. secessionist states
   d. satellite states
   e. rump or truncated states

7. Boundaries that are based on neither cultural or physical features are often:
   a. ethnographic
   b. geometric
   c. natural
   d. irregular
   e. none of the above

8. An excellent example of a relic boundary exists:
   a. between Canada and Alaska
   b. between Brazil and Argentina
   c. between Laos and Cambodia
   d. within Germany
   e. within Ireland

9. A federal government is usually considered a less geographically expressive system.
   a. true
   b. false

10. The United States, Canada, France, and Australia are all examples of federal governments.
    a. true
    b. false

11. Whatever disrupts internal order and encourages destruction of the country is called:
    a. *raison d'etre*
    b. insurgency
    c. centrifugal force
    d. nationalism
    e. centripetal force

12. Electoral geography is useful for identifying:
    a. the Lower South
    b. formal and functional culture regions
    c. spatial patterns of ethnicity
    d. supranational organizations
    e. none of the above

13. The creation of districts that have a majority of voters favoring the group in power and a minority of opposition voters is called:
   a. gerrymandering
   b. cleavage control districts
   c. redistricting
   d. *raison d'etre*
   e. selective electoral districts

14. During political diffusion, the original core area seldom remains the country's most important district.
   a. true
   b. false

15. The Russian state originated in the small principality of:
   a. St. Petersburg
   b. Kiev
   c. Warsaw
   d. Novgorad
   e. Moscow

16. Potentially, countries without political core areas, such as Zaire and Belgium, are the least stable of all independent states.
   a. true
   b. false

17. The best example of contagious expansion diffusion in political geography is:
   a. the growth of the Russian Empire
   b. political independence in Africa
   c. the creation of Mexico
   d. British colonization of India
   e. the reunification of Germany

18. Throughout much of history, a country's survival was enhanced by some sort of natural protection. These areas of protection are called:
   a. national moats
   b. natural defenses
   c. ethnic islands
   d. national shields
   e. folk fortresses

19. The "heartland theory" was developed by geographer:
   a. Wilbur Zelinsky
   b. Frank Ainsley
   c. Halford Mackinder
   d. Jean Gottmann
   e. Derwent Whittlesey

20. The "heartland theory" predicted, in effect:
    a. the growth of Southern Democrats
    b. the economic power of the Midwest
    c. the rise of French and British colonialism
    d. Russian conquest of the world
    e. communism in mainland China

21. Which of the following is <u>not</u> an example of a modern nation-state?
    a. Sweden
    b. Belgium
    c. Japan
    d. Armenia
    e. Germany

22. The greatest concentration of the francophone cultural-linguistic minority in Canada is found in:
    a. Toronto
    b. Ontario
    c. Manitoba
    d. Newfoundland and New Brunswick
    e. Quebec

23. The "impress of central authority" refers to:
    a. usually unitary rather than federal states
    b. transportation networks
    c. omnipresent military authority
    d. government and landscape
    e. none of the above

24. A good example of national iconography in the American landscape is:
    a. bald eagles and flags
    b. the National Cathedral
    c. the Grand Canyon
    d. New York City
    e. baseball

25. This former principality is used as an example for illustrating the relationship between terrain and political geography:
    a. Andorra
    b. Liechtenstein
    c. Bavaria
    d. Berchtesgaden
    e. Denmark

PART TWO: Short answer (probable essay type questions)

26. Briefly describe the meaning of the "territorial imperative":

27. How and why are political boundaries barriers to cultural diffusion?

28. State some of the contributions of geographer Halford Mackinder:

29. Briefly explain the process of contagious expansion diffusion by using the example of political independence in Africa:

30. What are folk fortresses and are they significant today?

31. Briefly summarize the importance of the heartland theory:

32. What are some of the environmental catastrophes associated with the Persian Gulf War?

33. What is the difference between a nation-state and a multinational country? Use examples from each.

34. Briefly explain the cleavage model in political geography:

35. How is the legal code imprinted on landscapes? Use examples.

36. Describe the various types of boundaries between and within countries.

37. What is some of the local political iconography of your state, province, or region?

38. Define the term "territoriality."

39. Provide some examples and explain the difference between exclaves and enclaves.

40. Briefly describe and explain the difference between unitary and federal spatial organization of government:

CHAPTER FIVE

THE MOSAIC OF LANGUAGES

Extended Chapter Outline (including Key Terms)

I. Linguistic Culture Regions
   A. Language Families
      1. Indo-European
      2. Afro-Asiatic
      3. Others
   B. English Dialects in the United States

II. Linguistic Diffusion
   A. Indo-European Diffusion
   B. Austronesian Diffusion
   C. Searching for the Primordial Tongue

III. Linguistic Ecology
   A. The Environment and Vocabulary
   B. The Environment Provides Refuge
   C. The Environment Guides Migration

IV. Culturo-Linguistic Integration
   A. Technology and Linguistic Dominance
   B. The Social Morale Model
   C. The Economic Development Model
   D. Language and Religion

V. Linguistic Landscapes
   A. Toponyms
   B. Generic Toponyms of the United States
   C. Toponyms and Cultures of the Past
   D. Toponyms and Environmental Modification

VI. Conclusion

LEARNING OBJECTIVES OF CHAPTER FIVE

After reading this chapter and studying the maps and illustrations, you should be able to:

1. Explain and understand the difference between a language family, a language, a dialect, and a linguistic accent.

2. Describe the diffusion process of several major language families throughout the world.

3. Demonstrate the various relationship between language and physical environment.

4. Discuss the influence of physical and cultural environment on languages.

5. Explain several models of cultural integration involving language.

6. Understand the role of language and religion.

7. Interpret the cultural landscape through the reading of various toponyms.

8. Grasp a better understanding of the place or region where you live by interpreting local toponyms.

SELECTED MAP READING AND INTERPRETATION

This section of the Study Guide is intended to heighten your map reading and interpretation skills. It will also help you apply the text readings to visual and spatial display of concepts, themes, and examples in cultural geography.

A world atlas will be very useful in completing this section of the Study Guide and will enhance your comprehension of the maps in the textbook. Ask your instructor to recommend an appropriate atlas to purchase (or visit the map collection at your library). A world atlas is essential for your personal reference library, not only during this course, but throughout your college career.

After reading the text and then studying the accompanying map and its captions, answer the following questions:

FIGURE 5.3 "Major linguistic culture regions of the world"

1. What is the most widespread or spatially dominant language sub-group spoken on each continent? (It is important to know the boundary between Asia and Europe!)

____

____

____

2. List five countries where the Austronesian language sub-group is spoken, including an African island state.

____

____

____

3. Looking at this map, do you see the reason behind the name "Indo-European" family? Explain:

____

____

____

4. Identify the language sub-group for the following independent countries: Vietnam, South Korea, Cameroon, Eritrea, Suriname, Hungary, Iceland, and Lebanon.

____

____

____

FIGURE 5.6 "Origin and diffusion of four major language families in the eastern hemisphere"

1. What present day countries comprise the source area or hearth of the Niger-Congo language family?

____

____

____

2. Where is the original source area of modern English?

____

____

____

3. Compare this map with Figure 3.12 in Chapter Three. Are there similar patterns of certain diffusion routes? If so, what conclusion can be drawn?

____

____

____

FIGURE 5.9 "The environment is linguistic refuge in the Caucasus Mountains"

1. In the 1990s we are witnessing a secessionist movement in the region of Chechnya. In which independent country is Chechnya located and what is the language family sub-group of the Chechen language?

____

____

____

2. Can you determine an east-west spatial pattern among two of the language family sub-groups? What is it?

____

____

____

3. Does this map show greater diversity of languages at higher or lower elevations of terrain? (You will probably require an atlas to find the answer.) Can you explain the basis for the spatial patterns?

____

____

____

FIGURE 5.15 "Generic place names . . . "

1. Study the map. Can you think of any reasons, historical or physical, why the areas of upstate New York and northern Maine are practically devoid of the place name types indicated on this map?

____

____

____

2. There is a concentration of typical place-name characteristics of New England found in northeastern Ohio. Can you think of any reasons for this concentration and why other areas of Ohio have far fewer of these toponyms?

____

____

____

FIGURE 5.18 "Arabic toponyms in Iberia"

1. Using this map, speculate concerning the direction of the Moorish invasion and retreat, the duration of Moorish Islamic rule in different parts of Iberia, and the main centers of former Moorish power:

____

____

____

2. Based on the text reading and using an atlas, list some present day place-names that you consider to be of Arabic origin:

____

____

____

3. Look at the place-names on a map of you home state or province. Are any toponyms derived from languages other than English? (Perhaps from Spanish, French, or Native American languages.) What do these place-names reveal to you about the history of these regions?

____

____

____

CREATE YOUR PERSONAL GLOSSARY OF KEY TERMS, PEOPLE, AND PLACES

In the space below, write a definition and provide an example of each key term that is sufficient for **your understanding**. It is an excellent study habit to organize your response in three parts:

1. a formal definition or identification from the textbook
2. a definition of the key term in **your own words**
3. an example to add greater understanding of the key term

KEY TERMS, PEOPLE, AND PLACES FROM CHAPTER FIVE:

1. linguistic geography

2. languages

3. dialects

4. pidgin language

5. lingua franca

6. isoglosses

7. linguistic culture regions

8. linguistic "islands"

9. language families

10. language subfamilies

11. Indo-European family

12. Afro-Asiatic family

13. polyglot

14. Niger-Congo family

15. Altaic family

16. Austronesian family

17. Sino-Tibetan family

18. Austro-Asiatic family

19. minor language families

20. "regional words"

21. "Black English"

22. linguistic diffusion

23. Indo-European diffusion

24. Austronesian diffusion

25. Polynesian people

26. Primordial tongues

27. Nostratic

28. Dene-Caucasian

29. linguistic ecology

30. linguistic refuge areas

31. linguistic shatter belts

32. Culturo-linguistic integration

33. linguistic dominance

34. Treaty of Tordesillas

35. Social Morale model

36. bilingualism

37. monoglots

38. Economic Development model

39. changeover model

40. linguistic landscapes

41. toponyms

42. generic and specific place-names

43. the suffix "weiler"

44. the prefix "guada"

45. the suffix "ley" or "leigh"

46. Maori place names

47. Hans Kurath

48. "loanwords"

49. North American dialects

50. Caucasus Mountains

51. retreat of the Welsh language

52. relict language border

____

____

____

53. Arabic toponyms in Iberia

____

____

____

CHAPTER FIVE REVIEW: Self-evaluation Tests

PART ONE: Multiple-choice

Circle the best answer for each question. When you are finished, read each question again with your selected answer. After you are satisfied with your practice test, use the Answer Key in the back of the Study Guide to check your responses.

1. Languages can be defined as:
   a. speech patterns of various ethnic groups
   b. tongues that can be mutually understood
   c. speech patterns of singular ethnic groups
   d. tongues that cannot be mutually understood
   e. a collection of recognizable and similar dialects

2. Approximately _______ languages are spoken in the world today.
   a. 850
   b. 1200
   c. 3000
   d. 4500
   e. 6000

3. One existing language may be elevated to the status of ____________, or language of communication and commerce, over a wide area where it is not the mother tongue.
   a. pidgin language
   b. pidgin dialect
   c. lingua franca
   d. official or national language
   e. none of the above

4. The borders of individual word usage or pronunciations are called:
   a. linguistic culture regions
   b. isoglosses
   c. language dependency zones
   d. language continuum
   e. linguistic islands

5. Languages in the Indo-European Language family do <u>not</u> include:
   a. Turkish
   b. German
   c. Farsi (Persian)
   d. English
   e. Romanian

6. Three major languages of the Semitic people are Hebrew, Amharic, and:
   a. Farsi (Persian)
   b. Syrian
   c. Turkish
   d. Arabic
   e. Greek

7. Swahili, an important language on East Africa, is a member of what language family?
   a. Altaic
   b. Austronesia
   c. Nilo-Saharan
   d. Niger-Congo
   e. Khoisan

8. The earliest speakers of Indo-European apparently lived in what is now:
   a. India
   b. Turkey
   c. Russia
   d. Germany
   e. the Fertile Crescent

9. The diffusion of which language family is strongly associated with island culture and vast expanses of ocean?
   a. Austronesian
   b. Indo-European
   c. Semitic
   d. Niger-Congo
   e. none of the above

10. The Spanish language, derived from Castile, as well as the Celtic tongues, are especially rich in words describing rough terrain, such as mountains.
    a. true
    b. false

11. A good example of a linguistic refuge area is:
    a. Poland
    b. Japan
    c. southern India
    d. South Africa
    e. the Caucasus region

12. Although physical barriers such as mountain ridges can slow groups from migrating from one place to another, they infrequently serve as linguistic borders.
    a. true
    b. false

13. The Treaty of Tordesillas divided ______________ between Portuguese and Spanish control.
    a. Mexico
    b. the Caribbean realm
    c. South America
    d. Brazil
    e. Argentina

14. The plight of the Welsh language in the United Kingdom illustrates which of Withers' models?
    a. social morale
    b. clearance
    c. economic development
    d. changeover
    e. all of the above

15. In Muslim lands such as parts of India, Bangladesh, and Indonesia, the language of religious ceremony is:
    a. the local language
    b. the regional lingua franca
    c. Dravidian, Bengali, and Malay
    d. Arabic
    e. Sanskrit

16. Another term for place names is:
    a. call signs
    b. landscape symbols
    c. toponyms
    d. signage
    e. none of the above

17. Many place-names consist of:
    a. generic parts
    b. symbolic parts
    c. specific parts
    d. both answers a & c
    e. none of the above

18. The place-name term "center" is frequently used in what American region?
    a. New England
    b. Deep South
    c. Upper Midwest
    d. California
    e. Southwest

19. East of the Elbe and Saale rivers, we find archaic place-names suffixes such as "ow, in," and "zig" (as in Berlin and Leipzig). Each of these suffixes denote a ___________ origin.
    a. German
    b. Teuton
    c. Polish
    d. Celtic
    e. Slavic

20. The remnants of descriptive Arabic place-names are commonly found in regions of:
    a. Italy
    b. Greece
    c. France
    d. Spain
    e. Hungary

21. It is a fact that language is the basis for the expression of all elements of culture.
    a. true
    b. false

22. This late centenarian was best known for his pioneer studies of the linguistic geography of the eastern United States:
    a. Alfred Weber
    b. Hans Kurath
    c. Karl Welte
    d. Alfred Kroeber
    e. James Cobban

PART TWO: Short answer (probable essay type questions)

23. Briefly list some of the major themes encompassed in the study of linguistic geography:

24. What defines a language family and what are some examples?

25. List some of the language subfamilies of the Indo-European language family:

26. Briefly trace the origin and diffusion of the Indo-European language family:

27. What are some of the barriers to language diffusion?

28. What are the major regional dialects of the United States? Do you agree with these designations? Why or why not?

29. Explain the diffusion of the languages of the Polynesian peoples, such as the Hawaiians, and why this diffusion merits special attention:

30. What are the principle language families of Africa and the Middle East?

31. Describe some of the relationships between the environment and vocabulary:

32. How does language guide migration? Is this phenomenon true today? Why or why not?

33. Provide some examples of both specific and generic toponyms:

34. What are some place names in the United States derived from Native American, French, and Spanish languages? Can you think of other toponym origins in your local region?

35. Briefly describe the various models of culturo-linguistic integration:

CHAPTER SIX

RELIGIOUS REALMS

Extended Chapter Outline (including Key Terms)

I. Religious Regions
   A. Christianity
      1. Western & Eastern
      2. Catholicism, Protestantism, Orthodoxy
   B. Islam
      1. Shiite
      2. Sunni
   C. Judaism
   D. Hinduism
      1. Jainism & Sikhism
   E. Buddhism
   F. Animism
   G. Secularization

II. Religious Diffusion
   A. The Semitic Religious Hearth
   B. The Indus-Ganges Hearth
   C. Barriers and Time-Distance Decay

III. Religious Ecology
   A. The Environment and Monotheism
   B. Religion and Environmental Modification
      1. teleological view
   C. Religion and Environmental Perception

IV. Cultural Integration in Religion
   A. Religion and Economy
      1. food and drink taboos
   B. Religious Pilgrimage
   C. Religion and Political Geography
      1. theocracy

V. Religious Landscapes
   A. Religious Structures
   B. Landscapes of the Dead
   C. Religious Names on the Land

VI. Conclusion

LEARNING OBJECTIVES OF CHAPTER SIX

After reading this chapter and studying the maps and illustrations, you should be able to:

1. Understand and describe the origin and spatial diffusion of major world religions and their subdivisions.

2. Explain why certain religions expanded over several continents, such as Islam, while other religions experienced little change in range or membership size.

3. Know and describe the difference between ethnic and proselytic religions. You should also know several examples from each type.

4. Understand the regional patterns of various branches of Christianity, especially in the United States.

5. Begin to understand the role of religion in shaping politics and the economy, especially in regards to territory.

6. Discuss the major theories explaining why three major monotheistic faiths, Islam, Judaism, and Christianity, began in the same geographic region, as well as the connections between these religions.

7. Understand the various relationships between the modification of the environment and different religions.

8. Discuss aspects and locations of animistic belief systems, including their attachment to the local natural world.

9. Interpret different burial systems and the resulting landscapes of the dead.

10. Begin to think about the role of religion in shaping the history, migration, settlement patterns, and cultural geography of your home region and perhaps your family and/or ancestors.

SELECTED MAP READING AND INTERPRETATION

This section of the Study Guide is intended to heighten your map reading and interpretation skills. It will also help you apply the text readings to visual and spatial display of concepts, themes, and examples in cultural geography.

A world atlas will be very useful in completing this section of the Study Guide and will enhance your comprehension of the maps in the textbook. Ask your instructor to recommend an appropriate atlas to purchase (or visit the map collection at your library). A world atlas is essential for your personal reference library, not only during this course, but throughout your college career.

After reading the text and then studying the accompanying map and its captions, answer the following questions:

<u>FIGURE 6.2 "The world distribution of major religions"</u>

1. Other than some African countries, where is animism evident in the world?

____

____

____

2. What are the primarily Protestant countries of Europe?

____

____

____

3. Is Hinduism limited to India? If not, where else is it found?

____

____

____

4. Describe the changes in major religions of Africa from north to south:

____

____

____

<u>FIGURE 6.3 "Distribution of religious groups in Lebanon"</u>

1. What are the patterns of religious groups in Lebanon?

____

____

____

2. Can you think of reasons why Lebanon serves as religious refuge area? What is the pattern of the physical terrain?

FIGURE 6.4 "Leading Christian denominations in the U.S. and Canada"

1. Can you explain the patterns of religion in Texas and Louisiana?

2. Why do you think that Lutheranism is most prevalent in the Upper Midwest?

3. Can you explain why some states, such as Ohio, have no denominational majority, while others such as Utah and Mississippi have basically only one majority?

4. Other than Catholic, Baptist, and Lutheran majority areas, what other spatial patterns are evident?

FIGURE 6.10 "The diffusion of Christianity in Europe, first to eleventh centuries"

1. Look at the patterns of diffusion by the year 300. In what way do these patterns suggest hierarchical expansion diffusion?

2. Compare this map to Figure 1.10. Who were the "knowers" (converts) and who were the laggards at this stage?

3. What barriers to diffusion might account for the uneven advance of Christianity by the year 1050?

____

____

____

4. What were the last "pagan" countries of Europe? They were not Christianized until the late fourteenth century.

____

____

____

<u>FIGURE 6.17 "Consumption and avoidance of pork are influenced by religion"</u>

1. Can you explain the pattern in the United States? Think of factors of economy and culture.

____

____

____

2. How do these global patterns correspond to the map (Figure 6.2) of major world religions?

____

____

____

CREATE YOUR PERSONAL GLOSSARY OF KEY TERMS, PEOPLE, AND PLACES

In the space below, write a definition and provide an example of each key term that is sufficient for **your understanding**. It is an excellent study habit to organize your response in three parts:

1. a formal definition or identification from the textbook
2. a definition of the key term in **your own words**
3. an example to add greater understanding of the key term

KEY TERMS, PEOPLE, AND PLACES FROM CHAPTER SIX:

1. religion

____
____
____

2. proselytic religions

____
____
____

3. ethnic religions

____
____
____

4. sacred space

____
____
____

5. spiritual geography

____
____
____

6. mystical places

____
____
____

7. monotheism and polytheism

8. Western Christianity

9. Eastern Christianity

10. Coptic Church

11. Maronites

12. Nestorians

13. Eastern Orthodoxy

14. Protestantism

15. "Bible Belt"

16. Islam

17. the Koran

18. Five Pillars of Islam

19. Shiite and Sunni Muslims

20. Judaism

21. Sephardim and Ashkenazim

22. Hinduism

23. Jainism

24. Sikhism

25. Buddhism

26. Confucianism

27. Taoism

28. Shintoism and Lamaism

29. Animism

30. secularization

31. Semitic religious hearth

32. ethnic religions

33. contact conversion

34. Indus-Ganges religious hearth

35. religious ecology

36. geomancy or "feng-shui"

37. Ellen Semple

38. teleological view

39. doctrine of ahimsa

40. food taboos

41. "dry" counties (USA)

42. pilgrimage

43. Lourdes, France

44. Mecca, Saudi Arabia

45. Ganges River

46. partition of India

47. theocracy

48. sacred landscapes

49. mosques and minarets

50. religious toponyms

CHAPTER SIX REVIEW: Self-evaluation Tests

PART ONE: Multiple-choice

Circle the best answer for each question. When you are finished, read each question again with your selected answer. After you are satisfied with your practice test, use the Answer Key in the back of the Study Guide to check your responses.

1. Religion can be defined as:
   a. cultural theory
   b. a set of beliefs
   c. worship of either one or many gods
   d. a way of life
   e. a form of organized cult

2. Religions that seek new members are termed:
   a. aggressive
   b. growth active
   c. proselytic
   d. charismatic
   e. ethnic religions

3. Judaism and Hinduism may be termed:
   a. aggressive
   b. growth active
   c. proselytic
   d. charismatic
   e. ethnic religions

4. An example of sacred space is:
   a. Dome of the Rock
   b. a Christian Church
   c. the Wailing Wall
   d. all of the above
   e. none of the above

5. Islam may be termed a(n) _____________ faith.
   a. polytheistic
   b. vernacular
   c. monotheistic
   d. ethnic
   e. none of the above

6. Christian Egyptians are often members of the Eastern group of Christianity called:
   a. the Coptic Church
   b. Maronites
   c. Eastern Orthodox
   d. Melkites
   e. Nestorians

7. Most of the Christians of the highland region of Lebanon are:
   a. Copts
   b. Nestorians
   c. Greek Orthodox
   d. Maronites
   e. Presbyterians

8. The core of the Mormon realm is:
   a. Nevada
   b. Missouri
   c. South Dakota
   d. North Dakota
   e. Utah

9. Which of the following is considered a prophet of Islam?
   a. Jesus
   b. Moses
   c. Muhammad
   d. Abraham
   e. all of the above

10. The stronghold of the Shiite branch of Islam is the country of:
    a. Iraq
    b. Iran
    c. Saudi Arabia
    d. Kuwait
    e. Libya

11. The Jews who eventually settled in central and eastern Europe were known as the:
    a. Ashkenazim
    b. Orthodox
    c. Hasidic
    d. Sephardim
    e. none of the above

12. The majority of the world's Jewish population lives in:
    a. Europe
    b. Israel
    c. Russia
    d. North America
    e. Poland and Romania

13. The concept of "ahimsa" is focused on the:
    a. caste system
    b. Dravidian ethnic divide
    c. idea of nirvana
    d. veneration of all life forms
    e. Five Pillars of Wisdom

14. Hinduism has splintered into diverse religious groups that are usually regarded as separate religions. Two major direct splinter groups are:
    a. Lamaism & Sikhism
    b. Taoism & Lamaism
    c. Coptic & Nestorian
    d. Sikhism & Jainism
    e. none of the above are correct

15. Buddhism is a religion derived from Taoism.
    a. true
    b. false

16. Lamaism prevails not only in Tibet, but also in:
    a. Nepal
    b. Mongolia
    c. China
    d. Thailand
    e. all of the above

17. People who are considered ______________ in their form of faith believe that rocks, rivers, and other natural features can possess spirits or souls.
    a. monotheistic
    b. theocratic
    c. animistic
    d. secularized
    e. heathen

18. The youngest religion of the Semitic Religious Hearth is:
   a. Islam
   b. Judaism
   c. Hinduism
   d. Christianity
   e. Sikhism

19. The use of missionaries primarily involves the concept of:
   a. hierarchical diffusion
   b. relocation diffusion
   c. distance-decay
   d. eminent domain
   e. all of the above

20. When political leaders (such as kings) were converted to Christianity and their subjects later followed, the process of religious diffusion was termed:
   a. hierarchical diffusion
   b. relocation diffusion
   c. distance-decay
   d. eminent domain
   e. all of the above

21. Buddhism, although strongly associated with lands and people of southeastern and eastern Asia such as China and Laos, actually began in the Indian subcontinent.
   a. true
   b. false

22. A good example of absorbing barriers in religious diffusion is the attempt to spread Christianity into China.
   a. true
   b. false

23. Chinese Buddhists originally invented the magnetic compass to serve the needs of reincarnation and the travel of the soul.
   a. true
   b. false

24. The geographer often associated with environmental determinism/influence on people and their religions is:
   a. Richard Hartshorne
   b. Yi-Fu Tuan
   c. Pradyumna P. Karan
   d. Ellen Semple
   e. Fred Kniffen

25. Many followers of Islam make religious pilgrimages to the holy cities of Mecca and:
   a. Jiddah
   b. Riyadh
   c. Damascus
   d. Cairo
   e. Medina

26. In India and elsewhere the word "ghat" refers to:
   a. any structure of sacred worship
   b. a sacred space reserved for the burying of the dead
   c. steps that often lead down to water
   d. pagan or heathen regions of "non-believers"
   e. a balance or harmony of human and nature

PART TWO: Short answer (probable essay type questions)

27. Provide examples of proselytic and ethnic religions and discuss the difference between them:

____

____

____

28. What are some of the general spatial patterns of religion in North America?

____

____

____

29. What are some of the characteristics of the Islamic faith?

____

____

____

30. What is polytheism and what are some examples of this form of religion?

____

____

____

31. Briefly discuss the origin and diffusion of Buddhism, including its fusion with native ethnic religions.

____

____

____

32. What are some common characteristics of animism and where can it be found today?

33. Briefly discuss theories regarding the origin of religions in the Semitic religious hearth:

34. Provide examples of relocation diffusion and hierarchical diffusion of religion:

35. What religions grew out of the Indus-Ganges religious hearth?

36. What are some absorbing barriers in the diffusion of religion?

37. What are the practices of "feng-shui" and how do they relate to the physical environment?

38. What is the religious significance of Mount Shasta in California?

39. How are environmental factors used to explain the origin of monotheistic faiths?

40. Explain the "teleological view":

41. How can religion influence people's perception of the environment?

42. Explain the use of fish and wine in Christianity as examples of the relationship between religion and economy:

43. What are the various food taboos among Hindus, Muslims, Mormons, and Jews?

44. What is the purpose of a religious pilgrimage and what are some very important pilgrimage sites for Jews, Muslims, and Christians?

45. What role did religion play in the 1947 partition of India?

46. How do cemeteries preserve truly ancient cultural traits? Provide some examples:

CHAPTER SEVEN

FOLK GEOGRAPHY

Extended Chapter Outline (including Key Terms)

I. Folk Culture Regions
  A. Material folk culture regions
  B. Folk food regions
  C. Folklore regions

II. Folk Cultural Diffusion
  A. Folk songs
  B. Agricultural fairs
    1. Fred B. Kniffen
  C. Hay stackers
  D. Blowguns

III. Folk Ecology
  A. Geophagy
  B. Folk medicine
  C. Environmental perception

IV. Cultural Integration in Folk Geography
  A. Mountain moonshine
  B. Country & Western music
    1. Bluegrass

V. Folk Landscapes
  A. Building materials
  B. Floor plan
  C. Irish folk houses
    1. Estyn Evans
  D. Folk housing in North America

VI. Conclusion

LEARNING OBJECTIVES OF CHAPTER SEVEN

After reading this chapter and studying the maps and illustrations, you should be able to:

1. Define, identify, and describe material folk culture and folklore regions.
2. Understand and explain various examples of folk culture diffusion.
3. Describe the relationship between the environment and folk cultures.
4. Provide examples and explain how various cultural practices are integrated in folk cultures.
5. Discuss how the cultural landscape is shaped and affected by folk cultures.
6. Recognize the role of folk culture in your own family, region, or national identity.

SELECTED MAP READING AND INTERPRETATION

This section of the Study Guide is intended to heighten your map reading and interpretation skills. It will also help you apply the text readings to visual and spatial display of concepts, themes, and examples in cultural geography.

A world atlas will be very useful in completing this section of the Study Guide and will enhance your comprehension of the maps in the textbook. Ask your instructor to recommend an appropriate atlas to purchase (or visit the map collection at your library). A world atlas is essential for your personal reference library, not only during this course, but throughout your college career.

After reading the text and then studying the accompanying map and its captions, answer the following questions:

FIGURE 7.1 "Folk cultural survival regions of the United States and Canada"

1. Identify the folk culture of your local region. Can you think of any examples of existing or former evidence of material folk culture?

____

____

____

2. From north to south, what are the folk culture regions of the Rocky Mountains?

____

____

____

3. What are the two locations of Acadian French folk culture? What is another name used for this culture in some locales?

____

____

____

FIGURE 7.6 "Switzerland: Where do newborn children come from?"

1. Where exactly would you draw the boundaries of the culture regions on this map?

____

____

____

2. How many culture regions can you designate on this map?

____

____

____

3. Using an atlas or other reference, determine if the culture regions on this map correspond to the language and ethnic identity of Switzerland.

FIGURE 7.11 "Former distribution of the blowgun . . . "

1. Were the two widely separated areas of blowgun use the result of independent invention or cultural diffusion?

2. Compare and contrast blowgun occurrence in the Indian and Pacific ocean countries to the distribution of the Austronesian languages (Figure 5.4):

3. Can you think of any reasons for the lack of blowgun use in continental Africa, Australia, or southern South America?

FIGURE 7.20 "Traditional building materials in folk architecture"

1. What factors may explain the general patterns on this map?

2. What role might the physical environment play in shaping the patterns shown in Europe?

3. Strongly settled and influenced by European immigrants, why doesn't the United States share similar traditional building materials with Europe?

CREATE YOUR PERSONAL GLOSSARY OF KEY TERMS, PEOPLE, AND PLACES

In the space below, write a definition and provide an example of each key term that is sufficient for **your understanding**. It is an excellent study habit to organize your response in three parts:

1. a formal definition or identification from the textibook
2. a definition of the key term in **your own words**
3. an example to add greater understanding of the key term

KEY TERMS, PEOPLE, AND PLACES FROM CHAPTER SEVEN:

1. folk culture

____

____

____

2. popular culture

____

____

____

3. material culture

____

____

____

4. nonmaterial culture

____

____

____

5. folklore

____

____

____

6. folk geography

____

____

____

7. material folk culture region

8. folk food region

9. folklore region

10. Scotch-Irish

11. folk songs

12. agricultural fairs

13. Fred Kniffen

14. beaverslide hay stacker

15. blowguns

16. folk ecology

17. geophagy

18. folk medicine

19. Upland Southern culture

20. curanderos

21. moonshine

22. stock-car racing

23. Country & Western music

24. Bluegrass music

25. folk architecture

26. traditional building materials

27. unit farmstead

28. courtyard farmstead

29. strewn farmstead

30. Irish folk houses

31. Estyn Evans

32. "New England large"

33. "upright and wing"

34. "saddlebag house"

35. "shotgun house"

36. "Ontario farmhouse"

37. "I-house"

CHAPTER SEVEN REVIEW: Self-evaluation Tests

PART ONE: Multiple-choice

Circle the best answer for each question. When you are finished, read each question again with your selected answer. After you are satisfied with your practice test, use the Answer Key in the back of the Study Guide to check your responses.

1. The best active example of a folk culture in the United States is the:
   a. Hispanic culture of the Southwest
   b. Cuban culture in Florida
   c. Amish culture in Ohio and Pennsylvania
   d. African-American culture in the Lower South
   e. family farm of the Midwest

2. Food, tools, furniture, buildings and clothing are all considered:
   a. material culture
   b. folklore
   c. living history
   d. folk geography
   e. none of the above

3. Folk dialects, religions, and worldviews can be regarded as:
   a. traditional lifestyles
   b. popular culture
   c. cultural artifacts
   d. nonmaterial culture
   e. none of the above

4. Active folk cultures would most likely be found in:
   a. North America
   b. Equatorial Africa
   c. Japan and Taiwan
   d. Scandinavia, especially Iceland
   e. the Middle East

5. Perhaps the most enduring feature of folk culture is:
   a. architecture
   b. language
   c. clothing
   d. occupation
   e. food

6. Folk culture spreads by the same processes of diffusion as do other elements and types of culture, but more rapidly.
   a. true
   b. false

7. The beaverslide hay stacker is a good example of material folk culture associated with:
   a. traditional Midwest family farms
   b. Mountain Western American culture
   c. Amish culture regions
   d. New England Yankee culture
   e. Upland South culture

8. If one or more nonfunctional features of blowguns occurred in both South America and Indonesia, then it is logical to surmise that the distribution of blowguns:
   a. is based on the theory of independent-invention
   b. is independent of functionality
   c. is a result of diffusion
   d. occurred after the year 1492
   e. none of the above

9. A subject of folk ecology studies, the deliberate eating of earth is called:
   a. alfisol digestion
   b. aridisol digestion
   c. ground-food nutrition
   d. geophagy
   e. "rock eating"

10. In the Mexican folk culture region along the southern border of Texas, folk medicine is still widely practiced by:
    a. curanderos
    b. shamans
    c. faith healers
    d. witch doctors
    e. caballeros

11. Moonshine corn whiskey has been associated with the folk culture of the ______________________ since the 1700s.
    a. Upper Midwest
    b. Gulf Coast
    c. Rocky Mountains
    d. Great Lakes
    e. Upland South

12. Country & Western music in the United States has its origins in the folk culture of the:
    a. Upper Midwest
    b. Gulf Coast
    c. Rocky Mountains
    d. Great Lakes
    e. Upland South

13. "Bluegrass" music has its beginnings in several states in the U.S., but derives its name from the state of:
    a. Texas
    b. Kentucky
    c. Virginia
    d. West Virginia
    e. Tennessee

14. Stones are often used as the traditional building material of the farmers of:
    a. Northern Europe
    b. Russia
    c. Canada
    d. the Mediterranean
    e. Southeast Asia

15. The Sarawak "longhouse" is a good example of traditional:
    a. stone construction
    b. half-timbering
    c. communal housing
    d. tropical nomadic architecture
    e. unit farmstead

16. A noted expert on Irish folk geography is:
    a. Fred Kniffen
    b. Estyn Evans
    c. Connor O'Laughlin
    d. Hubert Wilhelm
    e. Carl Sauer

17. An example of an African-American folk dwelling is the:
    a. shotgun house
    b. New England large
    c. upright and wing
    d. traditional Georgian
    e. saddlebag house

18. As Yankee folk migrated westward, they developed this type of house:
    a. shotgun house
    b. New England large
    c. upright and wing
    d. traditional Georgian
    e. saddlebag house

PART TWO: Short answer (probable essay type questions)

19. What are the main differences between folk and popular culture?

____

____

____

20. In which major world regions are active folk cultures most likely to be found?

____

____

____

21. Provide some examples of both material and nonmaterial culture:

____

____

____

22. What are some of the traditional folk culture regions of the western United States?

23. Discuss some culture elements and provide an example of a folk food region:

24. Use folk songs to demonstrate the process of folk cultural diffusion:

25. Briefly explain the two conflicting ideas or theories concerning the global distribution of the blowgun.

26. What is the reason behind the cultural practice of geophagy?

27. What are three major contributions to the folk geography of the United States attributed to the Upland South region?

28. Describe the theme of cultural integration in regards to the exchange between popular and folk culture:

29. Explain the development of stock-car racing in the United States in terms of folk culture:

30. Select an area of the world outside of North America and describe the traditional building materials and their relationship to the natural environment:

____

____

____

____

____

____

CHAPTER EIGHT

POPULAR CULTURE

Extended Chapter Outline (including Key Terms)

I. Popular Culture Regions
   A. Placelessness or Clustering?
   B. Cyberspace
   C. Food and Drink
   D. Popular Music
   E. Sports
   F. Beauty Pageants
   G. Vernacular Culture Regions

II. Diffusion in Popular Culture
   A. Advertising
   B. Diffusion of Classical Place-Names
   C. Diffusion of the Rodeo
   D. International Diffusion
   E. Communications Barriers

III. The Ecology of Popular Culture
   A. Environmental Influence
   B. Impact on the Environment

IV. Cultural Integration in Popular Culture
   A. The Convergence Hypothesis
   B. Mapping Personal Preference
   C. Place Images
   D. Social Spatialization

V. Landscapes of Popular Culture
   A. Elitist Landscapes
   B. Landscapes of Consumption
   C. Leisure Landscapes
   D. The American Scene

VI. Conclusion

LEARNING OBJECTIVES OF CHAPTER EIGHT

After reading this chapter <u>and</u> studying the maps and illustrations, you should be able to:

1. Recognize and describe the difference between folk culture and popular culture.
2. Identify and provide examples of popular life-style clusters.
3. Describe and understand formal culture regions in the United States.
4. Explain the diffusion process of classic place names and the rodeo in the United States.
5. Understand cultural integration in popular culture, including the significance of place images.
6. Understand the relationship between popular culture formal regions and social spatialization.
7. Recognize the often heavy impact of popular culture on the environment.
8. Identify and describe various landscapes of popular culture.
9. Begin to understand and contemplate the impact of popular culture on your own life-style, thought, and action.

SELECTED MAP READING AND INTERPRETATION

This section of the Study Guide is intended to heighten your map reading and interpretation skills. It will also help you apply the text readings to visual and spatial display of concepts, themes, and examples in cultural geography.

A world atlas will be very useful in completing this section of the Study Guide and will enhance your comprehension of the maps in the textbook. Ask your instructor to recommend an appropriate atlas to purchase (or visit the map collection at your library). A world atlas is essential for your personal reference library, not only during this course, but throughout your college career.

After reading the text and then studying the accompanying map and its captions, answer the following questions:

FIGURE 8.4 "Three examples of the forty 'life-style clusters' in U.S. popular culture"

1. How was the data collected for the creation of this map?

____

____

____

2. What are the definitions of "Gray Power," "Old Yankee Rows," and "Norma-Rae-ville"?

____

____

____

3. How do you see this map changing during the next twenty years?

____

____

____

FIGURE 8.6 "The cultural geography of commercial beer consumption"

1. Do you think this pattern of beer consumption is influenced by the partial German ancestry of nearly one-fourth of all Americans? Why or why not?

____

____

____

2. What cultural elements might be responsible for the pattern of low beer consumption in the Lower South?

____

____

____

3. Do you think the patterns on this map have changed since the data represent consumption in 1972? How and why?

FIGURE 8.7 "Fast-food sales as a share of total restaurant sales, by state":

1. What might account for the spatial variation in this aspect of popular culture?

2. Does this pattern bear any similarity to the map of traditional and folk cultures shown in Figures 1.6 and 7.1?

3. What does this suggest about the convergence hypothesis, which holds that regional cultures in America are collapsing into a national culture?

FIGURE 8.14 "The vernacular 'Middle West' or 'Midwest'"

1. What might account for the differences between this map and Figure 8.13?

2. Why would some respondents (less than 10 percent) claim to live in the Midwest in states such as Florida, California, and New Hampshire?

3. Which states on this map, such as Ohio, can be considered on the Midwest periphery?

FIGURE 8.28 "The gentleman farm is an elitist landscape feature in central Kentucky"

1. According to Karl Raitz, what are the characteristics of this landscape and why did it develop in this location?

____

____

____

2. How does such a landscape differ from one that might be produced by a folk culture?

____

____

____

3. Considering this landscape, what problems do you foresee for the spatial growth of the city of Lexington?

____

____

____

____

CREATE YOUR PERSONAL GLOSSARY OF KEY TERMS, PEOPLE, AND PLACES

In the space below, write a definition and provide an example of each key term that is sufficient for **your understanding**. It is an excellent study habit to organize your response in three parts:

1. a formal definition or identification from the textbook
2. a definition of the key term in **your own words**
3. an example to add greater understanding of the key term

KEY TERMS, PEOPLE, AND PLACES FROM CHAPTER EIGHT:

1. popular culture

___

___

___

2. "growth, progress, fad, and trend"

___

___

___

3. placelessness

___

___

___

4. clustering

___

___

___

5. cyberspace

___

___

___

6. food and drink in popular culture

___

___

___

7. popular music

8. sports and popular culture

9. beauty pageants

10. vernacular culture

11. vernacular regions

12. Wilbur Zelinsky

13. reverse hierarchical diffusion

14. classical place-names

15. "classical belt"

16. rodeo

17. international diffusion

18. geography of rock and roll

19. spelunkers

20. the convergence hypothesis

21. mapping personal preference

22. place image

23. social spatialization

24. elitist landscape

25. gentleman farm

26. Kentucky Bluegrass Basin

27. landscape of consumption

28. West Edmonton Mall

29. Karl Raitz

30. landscape of leisure

31. the "American Scene"

32. residential front yards

33. desert front yards

34. sense of place

____
____
____

35. fast-food

____
____
____

36. sport regions

____
____
____

37. the vernacular "Midwest"

____
____
____

CHAPTER EIGHT REVIEW: Self-evaluation Tests

PART ONE: Multiple-choice

Circle the best answer for each question. When you are finished, read each question again with your selected answer. After you are satisfied with your practice test, use the Answer Key in the back of the Study Guide to check your responses.

1. A spatial standardization that diminishes cultural variety and demeans the human spirit can result in:
   a. place image
   b. non-place image
   c. placelessness
   d. aspatial values
   e. popular culture

2. According to the text, the leading region or places for beer consumption in the U.S. is:
   a. Lower South
   b. Upper Midwest
   c. California and Florida
   d. Northeast
   e. Southwest

3. Nearly all professional sports in the U.S. were first organized in the:
   a. 1920s
   b. 1950s
   c. 1930s
   d. 1890s
   e. nineteenth century

4. Regions perceived to exist by their inhabitants are termed:
   a. popular regions
   b. vernacular regions
   c. folk regions
   d. perceived regions
   e. psychological regions

5. The example of Wal-Mart is used illustrate the concept of:
   a. reverse hierarchical diffusion
   b. space-time distance decay
   c. popular marketing diffusion
   d. placelessness
   e. none of the above

6. Time-distance decay in popular culture diffusion is ____________ than/as folk culture.
   a. stronger
   b. considerably stronger
   c. weaker
   d. considerably weaker
   e. about the same

7. Popular culture is considered to have a ________ rate of diffusion.
   a. slow
   b. average
   c. rapid
   d. lateral
   e. dynamic

8. The rodeo, as an example of popular culture diffusion, has its origins in:
   a. Spanish settlement areas
   b. primarily Texas
   c. Montana and Oklahoma
   d. southern California
   e. both answers b & c

9. This geographer, professor emeritus at Penn State, is best known for his contributions to the study of American popular culture:
   a. Ronald Isaac
   b. Thomas Leinbach
   c. John Fraser Hart
   d. Frank E. Bernard
   e. Wilbur Zelinsky

10. Of the following European countries, which probably do not participate much in Western popular culture?
    a. Czech Republic, Lithuania, Austria
    b. Switzerland, France, Luxembourg
    c. Norway, Finland, Estonia
    d. Germany, Poland, Belgium
    e. none of the above

11. Popular culture, by definition, does not display regionalization.
    a. true
    b. false

12. The devices of diffusion in popular culture require large amounts of electricity and fossil fuels.
    a. true
    b. false

13. Popular culture makes exceedingly high demands on:
    a. National Parks
    b. wilderness areas
    c. coastal zones
    d. National Recreation areas
    e. all of the above

14. Many social scientists assume that the results of causal forces in popular culture would be to homogenize culture. This view is called:
    a. placelessness
    b. time-distance continuum
    c. clustering
    d. convergence hypothesis
    e. none of the above

15. Photography, literature, television, and film often contribute to the creation of:
    a. peripheral place zones
    b. tourist illiteracy
    c. place image
    d. the modernity hypothesis
    e. none of the above

16. The geographer associated with the cultural geography of the Kentucky Bluegrass Basin is:
   a. Wilbur Zelinsky
   b. Bill Berentsen
   c. Karl Raitz
   d. Vaclav Svambera
   e. Anne Verlent

17. Which of the following would probably be labeled as a gentleman farm?
   a. sheep ranching for wool production
   b. horticulture of nuts and fruits
   c. fish farming in the Mississippi
   d. thoroughbred horse breeding in Kentucky
   e. organic farming

18. In stage 5 of Jakle and Mattson's model of commercial strip evolution, the:
   a. residential function disappears
   b. commercial function dominates
   c. gas stations are introduced
   d. single-family residence dominates
   e. none of the above

19. The most reflective landscape of consumption in popular culture is the:
   a. commercial strip
   b. shopping mall
   c. central business district (CBD)
   d. gentleman farm
   e. hotel and convention complexes

20. Lowenthal concludes that the preeminence of ________________ is an important part of the "American Scene."
   a. cult of bigness
   b. ugliness
   c. big hair and TV talk shows
   d. function over form
   e. manicured front lawns

PART TWO: Short answer (probable essay type questions)

21. Briefly describe the difference between popular and folk culture:

____

____

____

22. List some of the defining elements of popular culture:

23. Provide an example and define "placelessness."

24. What is the role of popular music in the geography of popular culture?

25. Discuss the regional variation of beauty pageants and rodeos in the United States:

26. What is a vernacular culture region and how do they apply to popular culture?

27. Discuss the role of advertising in the process of diffusion:

28. What are some barriers to the diffusion of popular culture?

29. What are landscapes of tragedy? Explain.

30. What are some of the environmental influences on popular culture?

31. State some specific examples of popular culture's impact on the environment:

32. Explain the convergence hypothesis. Do you agree with it?

33. What is a place image and how are place images created?

34. Provide some characteristics of an elitist landscape:

35. What is a landscape of consumption?

36. Provide a definition and example of a leisure landscape:

37. Explain the meaning of Lowenthal's "The American Scene":

38. What are some examples of popular culture in your home area?

39. Why do you think American pop culture has become an international phenomenon?

40. As North Americans, what features of other countries' popular culture do we absorb?

____

____

____

CHAPTER NINE

ETHNIC GEOGRAPHY

Extended Chapter Outline (including Key Terms)

I. Ethnic Regions
  A. Ethnic Culture Regions in Rural North America
  B. Urban Ethnic Neighborhoods and Ghettos
  C. Other Ethnic Migrants
  D. Ethnic Mix and National Character

II. Cultural Diffusion and Ethnicity
  A. Migration and Ethnicity
  B. Simplification and Isolation

III. Ethnic Ecology
  A. Cultural Preadaptation
  B. Ethnic Environmental Perception
  C. Ecology of Ethnic Survival

IV. Ethnic Cultural Integration
  A. Ethnicity and Business Activity
  B. Ethnicity and Type of Employment
  C. Ethnicity and Farming Practices

V. Ethnic Landscapes
  A. An "Ethnic Flag"
  B. Ethnic Settlement Patterns
  C. Urban Ethnic Landscapes

VI. Conclusion

LEARNING OBJECTIVES OF CHAPTER NINE

After reading this chapter <u>and</u> studying the maps and illustrations, you should be able to:

1. Present examples and explain how various groups define "ethnicity."
2. Understand the definitions of, and relationships between, national character, national origin, "nationality," race and ethnicity.
3. Discuss the process and characteristics of ethnic migration.
4. Discuss ethnic settlement patterns, cultural preadaptation, and ethnic survival.
5. Describe the cultural integration of ethnicity and business, employment, and farming.
6. Identify and describe elements of the ethnic landscape, including settlement patterns and urban landscapes.
7. Distinguish between an ethnic homeland and an ethnic island.
8. Understand the processes involving ethnicity and assimilation, acculturation, and re-awareness.
9. Use ethnic geography as a method of understanding the process of relocation diffusion.
10. Begin to understand the role of ethnicity in your family, ancestors, and local neighborhood or region.

SELECTED MAP READING AND INTERPRETATION

This section of the Study Guide is intended to heighten your map reading and interpretation skills. It will also help you apply the text readings to visual and spatial display of concepts, themes, and examples in cultural geography.

A world atlas will be very useful in completing this section of the Study Guide and will enhance your comprehension of the maps in the textbook. Ask your instructor to recommend an appropriate atlas to purchase (or visit the map collection at your library). A world atlas is essential for your personal reference library, not only during this course, but throughout your college career.

After reading the text and then studying the accompanying map and its captions, answer the following questions:

FIGURE 9.2 "Ethnic minorities in China"

1. Which of the ethnic regions are homelands and which are islands?

___

___

___

2. See Figure 2.1. Why are China's ethnic groups concentrated in sparsely populated peripheries of the country?

___

___

___

3. Can you predict a break-up of the People's Republic of China, similar to the collapse and partition of the Soviet Union? Based on ethnicity, what new countries would you foresee?

___

___

___

FIGURE 9.4 "Ethnic and national-origin groups in North America"

1. What is the specific Slavic ethnic group shown in Texas?

___

___

___

2. Where are the areas of Dutch ethnicity on the map?

___

___

___

3. Which state shows only "English" ethnicity?

4. Is there a relationship between the African ethnic area and the terrain or physical environment?

FIGURE 9.5 "Selected ethnic homelands in North America . . . "

1. What are the viable ethnic homelands and why do you think they are considered viable?

2. What are some of the ethnic islands in the regions of ethnic island concentrations?

3. Because they are left blank, are states such as Kentucky without ethnicity? Why or why not?

4. Why are there no ethnic homelands indicated for Native Americans?

FIGURE 9.16 "The cultural ecology of Czech farm settlements in Texas"

1. What is the relationship between Czech settlement and prairie grasses?

2. Does it surprise you that a large Czech settlement exists in Texas? Why do you think Czechs settled here? Is it an example of preadaptation? Is the climate and soil similar to their native homeland?

FIGURE 9.22 "Dominant ethnic restaurant cuisine in North America"

1. Why do Mexican "foodways" balloon northward faster than the main diffusion of these people?

___

___

___

2. Why does Italian cuisine prevail so much farther south of the concentration of Italian ethnicity?

___

___

___

3. Do you feel this map is an accurate representation of ethnic cuisine in America? Why or why not?

___

___

___

CREATE YOUR PERSONAL GLOSSARY OF KEY TERMS, PEOPLE, AND PLACES

In the space below, write a definition and provide an example of each key term that is sufficient for **your understanding**. It is an excellent study habit to organize your response in three parts:

1. a formal definition or identification from the textbook
2. a definition of the key term in **your own words**
3. an example to add greater understanding of the key term

KEY TERMS, PEOPLE, AND PLACES FROM CHAPTER NINE:

1. "The Czech Capital of Nebraska"

2. ethnic group

3. host culture

4. acculturation

5. assimilation

6. ethnic geography

7. Walter Kollmorgen

8. ethnic minorities

9. ethnic homelands

10. ethnic islands

11. Acadiana

12. Hispano

13. Deseret

14. Black Belt

15. ethnic substrate

16. ethnic neighborhood

17. ghetto

18. nationality

19. chain migration

20. American letters

21. channelized

22. ethnic cleansing

23. return migration

24. cultural simplification

25. cultural adaptation

26. adaptive strategy

27. preadaptation

28. first effective settlement

29. Russian-Germans

30. aspen belt

31. tropical savanna climate

32. cultural maladaptation

33. Sami

34. economic performance

35. tortillerias

36. Dutch Calvinists

37. Hmong people

38. ethnic landscape

39. "ethnic flag"

40. matrilocally

41. ethnic turf

42. Old World

43. Walter Kollmorgen

____

____

____

44. Mennonite

____

____

____

45. ethnic cuisine

____

____

____

46. Little Havana

____

____

____

CHAPTER NINE REVIEW: Self-evaluation Tests

PART ONE: Multiple-choice

Circle the best answer for each question. When you are finished, read each question again with your selected answer. After you are satisfied with your practice test, use the Answer Key in the back of the Study Guide to check your responses.

1. The larger society in which an ethnic group resides is referred to as:
   a. ethnic island
   b. ethnic majority
   c. host culture
   d. national majority
   e. none of the above

2. When an ethnic group adopts enough of the ways of the larger society to function, this is termed:
   a. assimilation
   b. acculturation
   c. cultural adaption
   d. preadaptation
   e. melting pot

3. A complete blending with the larger society by an ethnic group is termed:
   a. assimilation
   b. acculturation
   c. cultural adaption
   d. preadaptation
   e. melting pot

4. The main difference between ethnic islands and ethnic homelands is:
   a. location
   b. spatial distribution
   c. level of cultural integration
   d. settlement patterns
   e. size

5. "Deseret" is a term used by some for the homeland of:
   a. Louisiana French
   b. French Canadians
   c. Mexican-Americans
   d. Jewish-Americans
   e. the Mormons

6. Ethnic islands are much more numerous than homelands but as common as ethnic substrates.
   a. true
   b. false

7. A ghetto is traditionally defined as a certain urban quarter where:
   a. people are a racial minority
   b. people are forced to live
   c. people of a minority religion live
   d. Jews traditionally live
   e. none of the above

8. The most numerous ethnic minorities in North American cities today are originally from:
   a. Africa
   b. India
   c. Mexico
   d. Asia
   e. former Communist lands

9. When compared to the United States, which of the following ethnic or national origin groups is poorly represented in Canada:
   a. African
   b. Hispanic
   c. German
   d. Mexican
   e. all of the above

10. According to the text and Table 9.1, the largest national origin/ethnic group in the United States is:
   a. Irish
   b. English
   c. German
   d. African
   e. Italian

11. In the country of Russia, only about 81 percent of the people are ethnically Russian.
   a. true
   b. false

12. About one-third of all Canadians claim single ancestry from this ancestry group:
   a. English
   b. French
   c. German
   d. Ukrainian
   e. Scottish

13. The decision for members of an ethnic group to migrate and the actual migration usually involves:
   a. hierarchical diffusion
   b. contagious diffusion
   c. relocation diffusion
   d. chain migration
   e. all of the above

14. When ethnic immigrants introduce their culture in a new land, a profound ______________ occurs.
   a. prejudice
   b. cultural simplification
   c. channelization
   d. return migration
   e. all of the above

15. The example of Finnish settlement patterns in Wisconsin represents a case of:
   a. preadaptation
   b. assimilation
   c. adaptive strategy
   d. first effective settlement
   e. none of the above

16. In America, the largest rural population of this ethnic group is found in Texas, where they were drawn to the ecological niche of tallgrass prairie soils.
    a. Germans
    b. Mexicans
    c. Swedes
    d. Scotch-Irish
    e. Czechs

17. The ecology of ethnic survival is often related to isolation and:
    a. language
    b. religion
    c. elevation
    d. food
    e. race

18. Light blue is a Greek ethnic color, but in Chinese urban neighborhoods the venerated and auspicious color is:
    a. black
    b. purple
    c. green
    d. red
    e. yellow

19. Green, a traditional color of ethnic Irish Catholics, is also found throughout the world in ________________ neighborhoods.
    a. Japanese
    b. Russian
    c. Muslim
    d. Indian
    e. French

20. Terminology in ethnic studies is often confusing. For example, if a native white South African migrates to the United States, does he become an "African-American?" In this and many respects, definitions of ethnic identity are often ones of perception and self-recognition.
    a. true
    b. false

PART TWO: Short answer (probable essay type questions)

21. What are some of the cultural and ethnic features that make Wilbur, Nebraska, "The Czech Capital of Nebraska"?

    ____
    ____
    ____

22. Define the terms ethnic group and host culture:

23. Explain the difference between acculturation and assimilation:

24. What is the focus of ethnic geography studies?

25. What are the two distinct geographical types of ethnic regions? Provide an example of each.

26. Provide an example of an ethnic homeland and an ethnic island in North America:

27. Explain the difference between an ethnic neighborhood and an ethnic ghetto:

28. Why are cities in North America more ethnically diverse than any other urban centers in the world?

29. What are the top five ethnic ancestry/national origin groups in the United States? How is this different from Canada?

30. Explain the concept of chain migration, including an example of an "American letter":

31. Briefly discuss the concept of preadaptation in ethnic migration:

32. Provide some examples of ethnic environmental perception:

33. What is the relationship between ethnicity and business activity?

34. What is the relationship between ethnicity and type of employment?

35. Who and where are the Hmong-Americans and what are their distinctive gardening practices?

36. Other than the Finnish sauna, what is an example of an "ethnic flag"?

37. How can colors connote and reveal ethnicity? Provide examples:

38. What are some of the contributions of Professor Walter Kollmorgen?

39. What are some local examples of ethnicity in your home town, city, or region?

40. Nearly everyone in North America is part of some ethnic fabric or identity. Describe the role of ethnicity or ethnic ancestry in your family or acquaintances.

CHAPTER TEN

THE CITY IN SPACE AND TIME

Extended Chapter Outline (including Key Terms)

I. Culture Region

II. Origin and Diffusion of the City
- A. Single-Factor and Multiple-Factor Models for the Rise of Cities
- B. Urban Hearth Areas
- C. Diffusion of the City from Hearth Areas

III. Evolution of Urban Landscapes
- A. The Greek City
- B. Roman Cities
- C. The Medieval City
- D. The Renaissance and Baroque Periods
- E. The Capitalist City
- F. Class, "Race," and Gender in the Industrial City
- G. Megalopolis
- H. Edge Cities
- I. The Urban Landscapes of the Developing World
- J. The Indigenous City
- K. The Colonial City
- L. The Emerging City

IV. The Ecology of Urban Location
- A. Site and Situation
- B. Defensive Sites
- C. Trade-Route Sites

V. Cultural Integration in Urban Geography

VI. Conclusion

LEARNING OBJECTIVES OF CHAPTER TEN

After reading this chapter <u>and</u> studying the maps and illustrations, you should be able to:

1. Define the "city" and discuss the meaning of urbanized population.
2. Explain various theories dealing with the origin and diffusion of the city.
3. Demonstrate a knowledge of urban hearth areas, and, the ten largest urban centers today.
4. Discuss the evolution of urban landscapes.
5. Understand the relationship between functional zonation and urban morphology.
6. Explain the causal factors that led to the capitalist city.
7. Discuss the impact of industrialization on city structure.
8. Understand various models of cities in developing countries.
9. Delineate the types of city sites and their functions.
10. Interpret with new understanding various cities with which you are familiar.

CREATE YOUR PERSONAL GLOSSARY OF KEY TERMS, PEOPLE, AND PLACES

In the space below, write a definition and provide an example of each key term that is sufficient for **your understanding**. It is an excellent study habit to organize your response in three parts:

1. a formal definition or identification from the textbook

2. a definition of the key term in **your own words**

3. an example to add greater understanding of the key term

KEY TERMS, PEOPLE, AND PLACES FROM CHAPTER TEN:

1. urbanized population

2. rural-to-city migration

3. World's 20 Largest Metropolitan Areas

4. primate city

5. single-factor model

6. hydraulic civilization

7. institution of kingship

8. multiple factor models

9. urban hearth areas

10. axis mundi

11. walled inner city

12. urban morphology

13. functional zonation

14. acropolis and agora

15. Etruscans

16. forum

17. Byzantine civilization

18. the medieval period

19. city "charter"

20. suburbs

21. Hangchow

22. cathedral

23. guildhall

24. Renaissance period

25. Baroque period

26. the capitalist city

27. industrial city

28. *laissez-faire* utilitarianism

29. benefits of agglomeration

30. zoning

31. "class" and "race" and "gender"

32. megalopolis

33. "edge cities"

34. Islamic world cities

35. the Colonial city

36. gridiron street pattern

37. barriadas

38. squatter settlements

39. "site" and "situation"

40. river-island site

41. acropolis sites

42. trade-route sites

43. spatial distribution of cities

44. "central-place" theory

45. tertiary stage of production

46. Walter Christaller

47. "threshold" and "range"

48. hinterlands

49. fifteenth-century humanism

50. Calcutta

CHAPTER TEN REVIEW: Self-evaluation Tests

PART ONE: Multiple-choice

Circle the best answer for each question. When you are finished, read each question again with your selected answer. After you are satisfied with your practice test, use the Answer Key in the back of the Study Guide to check your responses.

1. Urbanized population refers to a country's:
   a. number of cities
   b. number of cities over 100,000 in population
   c. percentage of urban population
   d. ratio of urban to rural population
   e. none of the above

2. The United States Census Bureau defines a city as a densely populated area of _______ people or more.
   a. 2,500
   b. 5,000
   c. 10,000
   d. 20,000
   e. 50,000

3. The term "world cities" refers to cities having populations of at least:
   a. 500,000
   b. 1,000,000
   c. 5,000,000
   d. 10,000,000
   e. none of the above

4. Which of the following cities is <u>not</u> ranked in the world's ten largest metropolitan areas?
   a. Seoul
   b. London
   c. Moscow
   d. New York
   e. Bombay

5. The city which dominates the political, economic, and cultural life of a country is the __________ city.
   a. capital
   b. most populous
   c. vanguard
   d. primate
   e. growth pole

6. The origin of cities is strongly related to:
   a. plant and animal domestication
   b. improved transportation networks
   c. warfare and defense
   d. improved building techniques
   e. all of the above

7. The hydraulic civilization model can be tied to all of the following urban hearths <u>except</u>:
   a. China
   b. Mesoamerica
   c. Egypt
   d. Mesopotamia (Iraq)
   e. Indus River valley

8. Cosmomagical cities exhibit the three spatial characteristics of a ________________, symbolic center, and cardinal direction orientation.
   a. walled circumference
   b. walled outer city
   c. universe-like form
   d. wheel spoke form
   e. moat or bulwark

9. The pattern of functional land use within a city is referred to as:
   a. urban morphology
   b. functional development
   c. functional zonation
   d. spatial landscapes
   e. urban land use

10. The place for public use and markets in ancient Greek cities was the:
   a. agora
   b. citadel
   c. forum
   d. acropolis
   e. castra

11. The _________ was the urban zone of a Roman city where religious, administrative, and educational structures, as well as markets, were located.
   a. agora
   b. citadel
   c. forum
   d. acropolis
   e. castra

12. The most important contribution or legacy of Roman city builders was probably:
    a. engineering
    b. transportation
    c. architecture
    d. site selection
    e. gridiron planning

13. The major functions of the medieval city are depicted in all of the following symbols <u>except</u> the:
    a. town hall
    b. charter
    c. fortress
    d. wall
    e. marketplace

14. The medieval town's crowning glory was usually the:
    a. town hall
    b. castle
    c. fortress
    d. cathedral
    e. clock tower

15. Most European cities were founded during the medieval period.
    a. true
    b. false

16. Few of the traditions of Western urban life began in the Medieval period.
    a. true
    b. false

17. The medieval landscape is still with us, giving us a visible history of the city and distinctive form.
    a. true
    b. false

18. The Baroque period predated the Renaissance in history.
    a. true
    b. false

19. The term and concept of megalopolis was created by:
    a. Walter Christaller
    b. Walter Kollmorgen
    c. Lewis Mumford
    d. Alfred Weber
    e. Jean Gottmann

20. A majority of the American people were not urbanized until:
   a. 1890
   b. 1920
   c. 1940
   d. 1950
   e. 1960

21. The regional setting of an urban location is called the:
   a. site
   b. node
   c. ecological niche
   d. situation
   e. none of the above

22. Mexico City, Montreal, New York City, and Venice are all examples of:
   a. limited access cities
   b. colonial cities
   c. river-island sites
   d. acropolis sites
   e. defensive sites

23. Critical to central-place theory is the fact that different goods and services vary both in ________ and range.
   a. spatial distribution
   b. quality
   c. quantity
   d. threshold
   e. access

24. An "edge city" is any place that has the following features, <u>except</u>:
   a. one million square feet or less of leasable office space
   b. 600,000 square feet or more of leasable retail space
   c. more jobs than bedrooms
   d. perceived by people as one place
   e. was not at all urban as recently as 30 years ago

25. Large scale squatter settlements are a typical feature of many:
   a. Mexican cities
   b. European and Russian cities
   c. "developing" world cities
   d. former French and Portuguese colonies
   e. North African and Middle East cities

PART TWO: Short answer (probable essay type questions)

26. Briefly explain "threshold" and "range" in central-place theory:

27. What did Christaller add to central-place theory in his second model?

28. List some common features of former colonial cities:

29. What are some of the major problems faced by cities in Africa, Mesoamerica, and India?

30. Describe a squatter settlement, including a couple of actual examples:

31. Clearly explain the difference between city site and situation:

32. Describe the alteration of Paris during the nineteenth century:

33. What is meant by "capitalist city"?

34. What role does "race" play in urban function and features?

35. Explain "*laissez-faire* utilitarianism" and its effects on cities:

36. Describe some main features of a medieval town:

37. What are the spatial features associated with cosmomagical cities?

38. Discuss an argument concerning the diffusion of the city from hearth areas:

39. What are the concepts of urban morphology and functional zonation?

40. What was a medieval town "charter" and what were its implications?

CHAPTER ELEVEN

THE URBAN MOSAIC

Extended Chapter Outline (including Key Terms)

I. Urban Culture Regions
   A. Social Regions
   B. Neighborhoods

II. Cultural Diffusion in the City
   A. Centralization
      1. Economic and social advantages
   B. Decentralization
      1. Economic and social advantages
      2. Public Policy
   C. The Costs of Decentralization
   D. Gentrification

III. The Cultural Ecology of the City
   A. The Urban Ecosystem
   B. The Urban Geologic Environment
   C. Urban Weather and Climate
   D. Urban Hydrology
   E. Urban Vegetation

IV. Cultural Integration and Models of the City
   A. Concentric Zone Model
   B. Sector Model
   C. Multiple Nuclei Model
   D. Feminist Critiques
   E. Apartheid and Postapartheid cities
   F. Soviet and Post-Soviet cities
   G. Latin American Model

V. Urban Landscapes
   A. Themes in Cityscape Study
      1. Landscape dynamics
      2. The City as Palimpsest
      3. Symbolic Cityscapes
      4. Perception of the City
   B. The New Urban Landscape

VI. Conclusion

LEARNING OBJECTIVES OF CHAPTER ELEVEN

After reading this chapter and studying the maps and illustrations, you should be able to:

1. Identify and describe various and distinct urban culture regions.

2. Understand the concept, views, perception, and importance of the "neighborhood" and its role in everyday life.

3. Explain the difference between and the pros and cons of centralization and decentralization of the city.

4. Understand the elements of the urban ecosystem and its impact on society.

5. Identify various city models and relate them to actual cities.

6. Understand the elements of the city landscape and its causal factors.

7. Identify symbolic, cultural, and perceptive elements of the city environment.

8. Begin to look and interpret, with new awareness, the cities and neighborhoods around you.

CREATE YOUR PERSONAL GLOSSARY OF KEY TERMS, PEOPLE, AND PLACES

In the space below, write a definition and provide an example of each key term that is sufficient for **your understanding**. It is an excellent study habit to organize your response in three parts:

1. a formal definition or identification from the textbook
2. a definition of the key term in **your own words**
3. an example to add greater understanding of the key term

KEY TERMS, PEOPLE, AND PLACES FROM CHAPTER ELEVEN:

1. social culture region

____

____

____

2. ethnic culture region

____

____

____

3. census tracts

____

____

____

4. "neighborhood"

____

____

____

5. definitions of homelessness

____

____

____

6. "zone in transition"

____

____

____

7. centralizing forces

8. decentralizing forces

9. economic and social advantages of centralization

10. agglomeration or clustering

11. "streetcar suburbs"

12. decentralization

13. economic and social advantages of decentralization

14. lateral commuting

15. Federal Housing Administration

16. covenants

17. red-lining

18. costs of decentralization

19. gentrification

20. sexuality and gentrification

21. costs of gentrification

22. urban ecosystem

23. urban geology

24. urban heat island

25. urban hydrology

26. runoff

27. urban vegetation

28. segregation

29. social group "invasion" and "succession"

30. concentric zone model

31. CBD

32. sector model

33. Homer Hoyt

34. Ernest Burgess

35. multiple-nuclei models

36. feminist critique

37. Latin American Model

38. spine/sector

39. Soviet and Post-Soviet cities

40. zone of accretion

41. zone of peripheral squatter settlements

42. urban landscape and cityscape

43. landscape dynamics

44. city as palimpsest

45. symbolic cityscape

46. ornamental iconography

47. perception of the city

48. landmarks

49. office parks

50. high-tech corridors

51. festival settings

52. militarized space

53. public space

CHAPTER ELEVEN REVIEW: Self-evaluation Tests

PART ONE: Multiple-choice

Circle the best answer for each question. When you are finished, read each question again with your selected answer. After you are satisfied with your practice test, use the Answer Key in the back of the Study Guide to check your responses.

1. An excellent method of defining social regions is to isolate one social trait and plot its distribution within a city by using:
   a. aerial maps
   b. satellite imagery
   c. census tracts
   d. telephone directories
   e. property tax records

2. The neighborhood concept is critical to cultural geography because it:
   a. recognizes the sentiment people have for places
   b. provides examples of race relations within cities
   c. illustrates and explores urban ecology
   d. helps us understand spatial structures of cities
   e. none of the above

3. An important economic advantage to central city location has traditionally been:
   a. product cost
   b. advertising
   c. close to major transport networks
   d. accessibility
   e. all of the above

4. Agglomeration or clustering is considered:
   a. an economic disadvantage
   b. an economic advantage
   c. the hope for the future of neighborhoods
   d. an out of date method of urban planning
   e. only used in North American cities

5. Today, many people travel to work from suburb to suburb. This is termed:
   a. suburban dependency
   b. "by-pass commuting"
   c. lateral commuting
   d. highway by-pass commuting
   e. none of the above

6. The government body known as the "FHA" is an abbreviation for:
   a. Fair Housing Act
   b. Federal Home Association
   c. Family Housing Act
   d. Federal Housing Act
   e. Federal Housing Administration

7. A practice in which banks and mortgage companies designate areas considered to be a high risk for loans is:
   a. redistricting
   b. red-lining
   c. covenants
   d. black listing
   e. none of the above

8. The movement of middle-class people into deteriorated areas of city center is referred to as:
   a. urban revitalization
   b. urban renewal
   c. gentrification
   d. insanity
   e. reverse out-migration

9. Lower-income people of the inner city are often displaced by the process of:
   a. urban sprawl
   b. urban housing projects
   c. gentrification
   d. homelessness
   e. reverse out-migration

10. The study of the relationship between an organism and its physical environment is called:
    a. physical geography
    b. physical geology
    c. Earth Science
    d. ecology
    e. cultural ecology

11. The direction of city growth, the patterning of social regions, and the routing of transportation can all be influenced by:
    a. topography
    b. river and waterway patterns
    c. climate
    d. seasonality
    e. temperature variation

12. The heat generation of a city produces a large mass of warmer air sitting over the city called the urban heat island.
    a. true
    b. false

13. During the summer, a city center is warmer than its suburbs.
    a. true
    b. false

14. The concentric zone model of Burgess has five zones. Zone 2, characterized by a mixed pattern of industrial and residential land use, is considered a ______________ zone.
    a. stable
    b. blue-collar
    c. transitional
    d. better housing
    e. gentrified

15. The area of the concentric zone model characterized by commuters and high-income families is zone ______ .
    a. 1
    b. 2
    c. 3
    d. 4
    e. 5

16. The sector model of urban land-use was developed by:
    a. Burgess
    b. Hoyt
    c. Harris
    d. Ullman
    e. none of the above

17. Freeways became part of the city landscape:
    a. after 1945
    b. before 1940
    c. since the 1950s
    d. only since the 1960s
    e. before 1930

18. The model that maintains that a city develops with equal intensity around various points is the __________________ model.
    a. sector
    b. concentric zone
    c. urban growth
    d. multiple nuclei
    e. none of the above

19. In the Latin American model the zone of maturity primarily includes:
    a. traditional colonial homes
    b. diverse housing types
    c. new industrialization
    d. squatter settlements
    e. business and commerce

20. A palimpsest is a:
    a. section of most American cities
    b. section of most European cities
    c. critical part of the sector model
    d. reference to transport planning
    e. parchment used over and over for written messages

21. A cemetery may be considered part of a city's
    a. park system
    b. symbolic landscape
    c. palimpsest
    d. CBD
    e. all of the above

22. Which of the following is considered a "landmark"?
    a. city hall
    b. railway depot
    c. industrial smokestack
    d. White Castle restaurant
    e. all of the above

23. Shopping malls in North America are ___________ spaces.
   a. eternal
   b. urban
   c. private
   d. open
   e. public

24. Mike Davis discusses patterns of "militarized" space. This space is primarily:
   a. military administration sections of cities
   b. military port urban zones
   c. any military base area, urban or rural
   d. defensive urban space
   e. none of the above

25. Many U.S. cities have a pattern of leap-frog expansion, and then in-filling, to use all of the space inside the city limits.
   a. true
   b. false

PART TWO: Short answer (probable essay type questions)

26. What human factors define a neighborhood?

____

____

____

27. What is meant by the "city as palimpsest"?

____

____

____

28. Describe elements of a neighborhood that you are familiar with:

____

____

____

29. What are the characteristics of the Latin American model zones?

____

____

____

30. What are the elements of Homer Hoyt's urban model?

____

____

____

31. How is the focus of this chapter different from the previous chapter?

32. List some symbols and landmarks of the urban environment:

33. How do cities affect the natural environment and ecology?

34. What does Kevin Lynch mean by "perception of the city"?

35. What are some criticisms of the concentric zone model?

36. What is the "feminist critique" in urban geography?

37. Briefly discuss the role of race and gender in the urban mosaic:

38. Describe the concept of the shopping mall as a social center:

39. What are some of the costs of city decentralization?

40. How did the streetcar and trolley affect neighborhood development in the late 19th and early 20th centuries?

41. Are neighborhoods found in the suburbs? Why or why not?

42. What are some of the human costs of gentrification?

43. How does topography influence city growth and structure?

44. Is urban hydrology a factor in the future of cities? How and why?

45. What are the problems of defining "homelessness"?

CHAPTER TWELVE

INDUSTRIAL GEOGRAPHY

Extended Chapter Outline (including Key Terms)

I. Industrial Regions
   A. Primary Industry
   B. Secondary Industry
   C. Tertiary Industry
   D. Quaternary Industry
   E. Quinary Industry

II. Diffusion of the Industrial Revolution
   A. Origins of the Industrial Revolution
   B. Diffusion from Britain

III. Industrial Ecology
   A. Radioactive Pollution
   B. Acid Rain
   C. The Greenhouse Effect and Ozone Depletion
   D. The "Green" Reaction
   E. Environmental Factors in Industrial Location

IV. Industrial Cultural Integration
   A. Labor Supply
   B. Markets
   C. The Political Element
   D. Industrialization and Cultural Change

V. Industrial Landscapes

VI. Conclusion

LEARNING OBJECTIVES OF CHAPTER TWELVE

After reading this chapter <u>and</u> studying the maps and illustrations, you should be able to:

1. Explain and define different categories of industry, including primary, secondary, tertiary, quaternary, and quinary.

2. Trace the locations, origin, and diffusion of the industrial revolution.

3. Understand the strong effects of industrialization upon the earth's environment and its people.

4. Explain the difference between ozone depletion and the Greenhouse Effect, as well as the consequences of each.

5. Discuss the causal factors in selecting and determining industrial location.

6. Identify elements of various industrial landscapes and their origin.

7. Contemplate the massive impact (evident today) of the Industrial Revolution on the earth, its people, and their cultures.

SELECTED MAP READING AND INTERPRETATION

This section of the Study Guide is intended to heighten your map reading and interpretation skills. It will also help you apply the text readings to visual and spatial display of concepts, themes, and examples in cultural geography.

A world atlas will be very useful in completing this section of the Study Guide and will enhance your comprehension of the maps in the textbook. Ask your instructor to recommend an appropriate atlas to purchase (or visit the map collection at your library). A world atlas is essential for your personal reference library, not only during this course, but throughout your college career.

After reading the text and then studying the accompanying map and its captions, answer the following questions:

FIGURE 12.2 "Regions of selected primary and secondary industries"

1. What major coastal regions show little or no commercial fishing activity? Can you think of any reasons for this?

____
____
____

2. What countries host the seven emerging major new centers of manufacturing?

____
____
____

3. Other than the U.S. and Russia, what countries have areas of lumbering and pulpwood?

____
____
____

4. What are the mining regions of Africa? What is being mined there?

____
____
____

FIGURE 12.3 "Major regions of industry in Anglo-America"

1. How would you delineate the major manufacturing belt in the U.S. and Canada?

____
____
____

2. Other than the major belt, what states share some of the minor regions? Why these locations?

FIGURE 12.12 "The diffusion of the railroad in Europe"

1. By 1845, what countries other than the United Kingdom, France, and Germany had railroads?

FIGURE 12.14 "The 500 worst counties of the U.S. in amount of toxic chemicals . . . "

1. What are the "ten worst states" on the Green Index?

2. Compare this map to Figure 12.3. What are your findings as a geographer?

3. Why are some of the worst counties located in non-industrial regions, such as northern Utah and Montana?

FIGURE 12.17 "Distribution of Acid Rain in North America"

1. Which states and provinces suffer the worst deposition problems? Why?

2. Why do you think there are problems of acid precipitation in the Pacific Northwest?

3. What is the ecology of these patterns? Include climate, weather, and topography in your answer.

____

____

____

CREATE YOUR PERSONAL GLOSSARY OF KEY TERMS, PEOPLE, AND PLACES

In the space below, write a definition and provide an example of each key term that is sufficient for **your understanding**. It is an excellent study habit to organize your response in three parts:

1. a formal definition or identification from the textbook
2. a definition of the key term in **your own words**
3. an example to add greater understanding of the key term

KEY TERMS, PEOPLE, AND PLACES FROM CHAPTER TWELVE:

1. Alfred Weber

____

____

____

2. industrial revolution

____

____

____

3. primary industry

____

____

____

4. secondary industry

____

____

____

5. tertiary industry

____

____

____

6. quaternary industry

____

____

____

7. quinary industry

8. renewable resources

9. nonrenewable resources

10. American Manufacturing Belt

11. economic core and periphery

12. uneven development

13. technopoles

14. deindustrialization

15. global corporations

16. postindustrial phase

17. multiplier leakage

18. high-tech corridors

19. silicon landscapes

20. cottage industry

21. guild industry

22. fossil fuels

23. textiles

24. metallurgy

25. mining

26. industrial ecology

27. radioactive pollution

28. Chernobyl

29. acid rain

30. acid precipitation

31. greenhouse effect

32. particulate pollutants

33. ozone layer

34. ozone depletion

35. "Green" reaction

36. "green" political parties

37. industrial inertia

38. energy supply

39. industrial cultural integration

40. labor-intensive industries

41. weight-gaining finished product

42. bulk-gaining finished product

43. import-export tariffs and quotas

44. NAFTA

45. industrial landscape

46. "Black Country"

47. topocide

48. distance in the preindustrial age

49. Wales

CHAPTER TWELVE REVIEW: Self-evaluation Tests

PART ONE: Multiple-choice

Circle the best answer for each question. When you are finished, read each question again with your selected answer. After you are satisfied with your practice test, use the Answer Key in the back of the Study Guide to check your responses.

1. The processing of raw materials into a more usable form is:
   a. primary industry
   b. secondary industry
   c. tertiary industry
   d. quaternary industry
   e. quinary industry

2. Agriculture is considered to be a:
   a. primary industry
   b. secondary industry
   c. tertiary industry
   d. quaternary industry
   e. quinary industry

3. The extraction of nonrenewable resources is a:
   a. primary industry
   b. secondary industry
   c. tertiary industry
   d. quaternary industry
   e. quinary industry

4. Legal services, retailing, and advertising are all considered:
   a. primary industry
   b. secondary industry
   c. tertiary industry
   d. quaternary industry
   e. quinary industry

5. Consumer related services such as education is considered a:
   a. primary industry
   b. secondary industry
   c. tertiary industry
   d. quaternary industry
   e. quinary industry

6. The American Manufacturing Belt is located:
    a. from Los Angeles to New York City
    b. primarily in the Lower South
    c. from Chicago to Pittsburgh
    d. around the Great Lakes and Northeast states
    e. New England and the Piedmont

7. Multinational or transnational companies are also referred to as:
    a. global corporations
    b. International enterprises
    c. Inter-regional industries
    d. global conglomerates
    e. none of the above

8. In Russia and Ukraine the most important mode of industrial transport is:
    a. highways
    b. railways
    c. airways
    d. waterways
    e. none of the above

9. While the cottage and guild industry systems were similar in many respects, only cottage industry depended on hand labor.
    a. true
    b. false

10. The initial breakthrough in the industrial revolution occurred in the:
    a. German steel industry
    b. British smelting processes
    c. Dutch papermaking
    d. British steel industry
    e. British textiles

11. The industrial revolution diffused from its place of origin first to:
    a. British colonies in North America
    b. British colonies in Africa and India
    c. continental Europe
    d. France
    e. West Midlands of England

12. The "Faustian bargain" refers to European dependency on:
    a. coal
    b. all fossil fuels
    c. petroleum
    d. nuclear power
    e. none of the above

13. The basic cause of acid rain is the burning of:
    a. wood
    b. coal
    c. oil
    d. fossil fuels
    e. high sulfur coal

14. Over 90 lakes are "dead" from acid rain in this seemingly pristine mountain range:
    a. Adirondacks
    b. Smoky Mountains
    c. Catskills
    d. Rockies
    e. Canadian Rockies

15. Greenhouse effect is caused primarily by:
    a. burning coal
    b. burning fossil fuels
    c. ozone depletion
    d. the hole in the ozone layer
    e. none of the above

16. Carbon dioxide gases permit solar shortwave radiation to reach the Earth's surface and allows outgoing long-wave radiation to escape.
    a. true
    b. false

17. Ozone layer depletion is caused by:
    a. fossil fuels
    b. greenhouse effect
    c. global warming
    d. manufactured chemicals
    e. radioactive waste

18. Industrial location theory was pioneered by:
    a. Carl Sauer
    b. Max Planck
    c. Alfred Weber
    d. Homer Hoyt
    e. Walter Christaller

19. In nearly all industrial site locations, a major factor is:
   a. water supply
   b. politics
   c. labor supply
   d. electricity supply
   e. climate

20. Coastal, often scenic, fishing villages from Norway to Portugal are part of the industrial landscape.
   a. true
   b. false

PART TWO: Short answer (probable essay type questions)

21. What and where is the origin of the Industrial Revolution?

____

____

____

22. What was the spatial-temporal pattern of industrial diffusion?

____

____

____

23. What are primary industrial activities?

____

____

____

24. What is the difference between quaternary and quinary industrial activities?

____

____

____

25. Where is most secondary industrial activity located in North America?

____

____

____

26. What was the diffusion pattern of the railroad in Europe?

____

____

____

27. Why did the highway displace the railroad in America as a major factor in regards to industry?

28. What are the sources of radioactive pollution?

29. How is acid rain created and what regions does it strongly affect?

30. What are the major problems associated with dependency on fossil fuels?

31. What are the main differences between greenhouse effect, ozone depletion, and global warming?

32. Explain the causal factors of greenhouse effect:

33. What is meant by the "green" reaction?

34. What are the pioneering contributions to industrial location theory advanced by Alfred Weber?

35. Describe some environmental factors in industrial location:

36. Describe the various relationships between labor and industrial location:

37. What all encompassing cultural changes were brought about by the industrial revolution?

38. Explain the various impacts politics can have on industry?

39. Describe industrial landscape from the categories of primary, secondary, and quaternary industries:

40. Provide examples of how the industrial landscape is interpreted by some humanists:

ANSWER SECTION

CHAPTER ONE: ANSWER KEY

MULTIPLE CHOICE: (textbook page references are in parentheses)

1.d (3) 2.b (4) 3.d (4) 4.a (4-5) 5.c (5) 6.d (4) 7.a (4) 8.c (7) 9.b (7-8) 10. a (8)

11.b (7) 12.a (10) 13.a (10) 14.c (10-11) 15.c (12) 16.e (27-28) 17.b (15) 18.c (15)

19.a (17) 20.d (21) 21.b (21) 22.a (22) 23.c (26) 24.e (27) 25.a (27-18) 26.b (31)

SHORT ANSWER:

There are often several acceptable answers to short answer questions. Provided below are the page numbers from the textbook where the information to answer the questions for this section of the self-evaluation test will be found.

27. title page
28. 3
29. 4-6
30. 5-7
31. 7
32. 7-8
33. 8-13
34. 10
35. 11
36. 8
37. 13-14
38. 15-17
39. 15
40. 17-21
41. 27-29
42. 23
43. 23-24
44. 25
45. 25-29
46. 26-27
47. 27
48. 27-31
49. 29-31

CHAPTER TWO: ANSWER KEY

MULTIPLE CHOICE: (textbook page references are in parentheses)

1.e (39) 2.c (39) 3.b (41) 4.a (41) 5.c (42) 6.b (42-43) 7.b (43) 8.a (41) 9.d (49-50)

10.c (44) 11.e (56) 12.d (51-52) 13.a (52) 14.a (53) 15.b (78-80) 16.e (56) 17.b (57)

18.c (61-62) 19.c (69) 20.e (72) 21.a (75) 22.b (78) 23.b (78) 24.d (83) 25.d (84)

SHORT ANSWER:

There are often several acceptable answers to short answer questions. Provided below are the page numbers from the textbook where the information to answer the questions for this section of the self-evaluation test will be found.

26. 39
27. 40-43
28. 42-43
29. 43-44
30. 48-50
31. 50-52
32. 52-53; 56
33. 52-53
34. 56-57
35. 59-60
36. 61-63
37. 64
38. 68-70
39. 70-71
40. 78-83

CHAPTER THREE: ANSWER KEY

MULTIPLE CHOICE: (textbook page references are in parentheses)

1.e (91) 2.a (91) 3.b (93) 4.d (95) 5.c (97) 6.b (113) 7.a (97) 8.d (97) 9.c (98)

10.e (100) 11.a (100) 12.c (101) 13.d (101) 14.b (102) 15.e (102) 16.e (104) 17.b (105)

18.b (107-108) 19.c (107) 20.a (108-110) 21.c (113) 22.c (116-117) 23.c (117)

24.a (114-118) 25.b (123)

SHORT ANSWER:

There are often several acceptable answers to short answer questions. Provided below are the page numbers from the textbook where the information to answer the questions for this section of the self-evaluation test will be found.

26. 123-124
27. 126
28. 127-130
29. 128-129
30. 110
31. 107-110
32. 94-96
33. 97, 119
34. 101-102
35. 103
36. 92-93
37. 106-111
38. 106-109
39. 114-115

CHAPTER FOUR: ANSWER KEY

MULTIPLE CHOICE: (textbook page references are in parentheses)

1.e (137) 2.a (137-138) 3.c (140-141) 4.b (140) 5.d (142) 6.b (142) 7.b (143)

8.d (143) 9.b (144) 10.b (144) 11.c (145) 12.b (147) 13.a (149-150) 14.b (150)

15.e (152) 16.b (152) 17.b (153) 18.e (157) 19.c (159) 20.d (159) 21.b (162)

22.e (166) 23.d (174) 24.a (175) 25.d (158)

SHORT ANSWER:

There are often several acceptable answers to short answer questions. Provided below are the page numbers from the textbook where the information to answer the questions for this section of the self-evaluation test will be found.

26. 140
27. 151
28. 159
29. 153-155
30. 157-159
31. 159-161
32. 161
33. 162-166
34. 166-168
35. 172
36. 172-174
37. 175
38. 138-139
39. 140-141
40. 143-144

CHAPTER FIVE: ANSWER KEY

MULTIPLE CHOICE: (textbook page references are in parentheses)

1.d (181) 2.e (181) 3.c (181) 4.b (182) 5.a (185) 6.d (186-187) 7.d (187)

8.b (191) 9.a (193) 10.a (195) 11.e (197) 12.b (199) 13.c (200) 14.e (201)

15.d (202-204) 16.c (206) 17.d (206) 18.a (206) 19.e (207-208) 20.d (208)

21.a (211) 22.b (188)

SHORT ANSWER:

There are often several acceptable answers to short answer questions. Provided below are the page numbers from the textbook where the information to answer the questions for this section of the self-evaluation test will be found.

23. 181
24. 184-185
25. 184-185
26. 191-193
27. 191-195
28. 188-191
29. 192-194
30. 185-187
31. 195-199
32. 199
33. 206-209
34. 206-209
35. 199-203

CHAPTER SIX: ANSWER KEY

MULTIPLE CHOICE: (textbook page references are in parentheses)

1.b (215) 2.c (216) 3.e (216) 4.d (225) 5.c (221) 6.a (217) 7.d (217) 8.e (219)

9.e (221) 10.b (222) 11.a (223) 12.d (223) 13.d (223) 14.b (223-224) 15.b (224)

16.b (224) 17.c (224) 18.a (228) 19.b (228) 20.a (228) 21.a (223) 22.b (231)

23.a (237) 24.d (237) 25.e (246) 26.c (251)

SHORT ANSWER:

There are often several acceptable answers to short answer questions. Provided below are the page numbers from the textbook where the information to answer the questions for this section of the self-evaluation test will be found.

26. 216-217
27. 218
28. 221-222
29. 223
30. 223-224
31. 224
32. 228-231
33. 228-231
34. 231
35. 231-232
36. 233-236
37. 235
38. 236-237
39. 238
40. 240-241
41. 241-242
42. 242-245
43. 245-246
44. 246-247
45. 251-252

CHAPTER SEVEN: ANSWER KEY

MULTIPLE CHOICE: (textbook page references are in parentheses)

1.c (259) 2.a (260) 3.d (260) 4.b (259) 5.e (262) 6.b (264) 7.b (266)

8.c (267-268) 9.d (271) 10.a (273) 11.e (276) 12.e (277-278) 13.b (278)

14.d (280-281) 15.d (283) 16.b (283) 17.a (285) 18.c (285)

SHORT ANSWER:

There are often several acceptable answers to short answer questions. Provided below are the page numbers from the textbook where the information to answer the questions for this section of the self-evaluation test will be found.

19. 259-260
20. 260-262
21. 260-264
22. 260
23. 262-263
24. 264-265
25. 267-269
26. 271-272
27. 274-279
28. 275-279
29. 275-277
30. 279-284

CHAPTER EIGHT: ANSWER KEY

MULTIPLE CHOICE: (textbook page references are in parentheses)

1.c (295) 2.b (301) 3.e (302-304) 4.b (305) 5.a (309) 6.d (309) 7.c (309)

8.a (312) 9.e (305) 10.e (308-313) 11.b (295-297) 12.a (315-317) 13.e (316-317)

14.d (318) 15.c (320) 16.c (323) 17.d (322) 18.a (324) 19.b (324) 20.d (326)

SHORT ANSWER:

There are often several acceptable answers to short answer questions. Provided below are the page numbers from the textbook where the information to answer the questions for this section of the self-evaluation test will be found.

21. 293-295 and Chapter Seven
22. 293-297
23. 295-296
24. 300
25. 305, 312-313
26. 305-308
27. 310
28. 308-313
29. 329-330
30. 314-317
31. 315-317
32. 318
33. 320
34. 321-324
35. 324-326
36. 326
37. 326-329
38. 293-331
39. 293-305
40. 293-331

CHAPTER NINE: ANSWER KEY

MULTIPLE CHOICE: (textbook page references are in parentheses)

1.c (337) 2.b (336) 3.a (336) 4.e (337) 5.e (339) 6.b (340) 7.b (434) 8.d (346)

9.e (348) 10.c (348) 11.a (348) 12.b (348) 13.e (349) 14.b (353) 15.a (364-365)

16.e (356-357) 17.c (357-358) 18.d (369) 19.c (370) 20.a (335-370)

SHORT ANSWER:

There are often several acceptable answers to short answer questions. Provided below are the page numbers from the textbook where the information to answer the questions for this section of the self-evaluation test will be found.

21. 335
22. 336
23. 336
24. 335-337
25. 337
26. 337-342
27. 343-345
28. 345-347
29. 348
30. 349-353
31. 354-357
32. 355-357
33. 358-361
34. 361-363
35. 363-364
36. 364-365
37. 368-370
38. 337
39. 335-370
40. 335-370

CHAPTER TEN: ANSWER KEY

MULTIPLE CHOICE: (textbook page references are in parentheses)

1.c (375) 2.a (376) 3.c (377) 4.b (377) 5.d (378) 6.a (378) 7.b (379) 8.c (382)

9.c (389) 10.a (390) 11.c (391) 12.d (393) 13.a (394-395) 14.d (397) 15.a (398)

16.b (398) 17.a (398) 18.b (398) 19.e (407) 20.b (402) 21.d (420-421) 22.c 423)

23.d (425-426) 24.a (409) 25.c (418)

SHORT ANSWER:

There are often several acceptable answers to short answer questions. Provided below are the page numbers from the textbook where the information to answer the questions for this section of the self-evaluation test will be found.

26. 425-428
27. 425-428
28. 411-417
29. 417-420
30. 417-420
31. 420-422
32. 399-400
33. 400-402
34. 402-405
35. 402-403
36. 394-398
37. 382-385
38. 386-388
39. 389
40. 394-395

CHAPTER ELEVEN: ANSWER KEY

MULTIPLE CHOICE: (textbook page references are in parentheses)

1.c (434) 2.a (435-436) 3.d (438) 4.b (439) 5.c (442) 6.e (443) 7.b (443)

8.c (445) 9.c (445) 10.d (449) 11.a (450-451) 12.a (452) 13.a (452) 14.c (456)

15.e (457) 16.b (458) 17.a (459) 18.d (459) 19.a (466) 20.e (468) 21.b (468)

22.e (470) 23.c (471-472) 24.d (477) 25.a (477)

SHORT ANSWER:

There are often several acceptable answers to short answer questions. Provided below are the page numbers from the textbook where the information to answer the questions for this section of the self-evaluation test will be found.

26. 435-436
27. 468
28. 435-436
29. 466-467
30. 458-459
31. 433-479
32. 468-469
33. 448-455
34. 469-470
35. 456-458
36. 460-462
37. 460-462
38. 471-472
39. 444-445
40. 441-442
41. 440-447
42. 447
43. 450-451
44. 453-454
45. 436-437

CHAPTER TWELVE: ANSWER KEY

MULTIPLE CHOICE: (textbook page references are in parentheses)

1.b (486) 2.a (487) 3.a (487) 4.d (493) 5.e (497) 6.d (490) 7.a (490) 8.b (493)

9.b (500) 10.e (501) 11.c (503) 12.d (505-507) 13.d (505) 14.a (506) 15.b (508)

16.b (508) 17.d (511) 18.c (515) 19.c (514) 20.a (520)

SHORT ANSWER:

There are often several acceptable answers to short answer questions. Provided below are the page numbers from the textbook where the information to answer the questions for this section of the self-evaluation test will be found.

21. 465-466
22. 464-468
23. 455
24. 463-464
25. 456
26. 468
27. 460-468
28. 470
29. 471-473
30. 473-475
31. 473-475
32. 473-475
33. 475
34. 475-476
35. 475-477
36. 477
37. 480-481
38. 478-480
39. 481-485
40. 481-485

END OF ANSWER KEY